Introduction to
SPECIAL
RELATIVITY

James H. Smith

Dover Publications, Inc.
Mineola, New York

Copyright

Copyright © 1965, 1993 by James H. Smith.
All rights reserved.

Bibliographical Note

This Dover edition, first published in 1995 and reissued in 2015, is an unabridged, unaltered republication of the work first published by W. A. Benjamin, Inc., New York, 1965.

Library of Congress Cataloging-in-Publication Data

Smith, James H. (James Hammond), 1925-
   Introduction to special relativity / James H. Smith. — Dover ed.
      p.   cm.
   Includes index.
   Originally published: New York : W.A. Benjamin, 1965.
   ISBN-13: 978-0-486-68895-4 (pbk.)
   ISBN-10: 0-486-68895-X (pbk.)
   1. Relativity (Physics)   2. Special relativity (Physics)   I. Title.
QC173.55.S63   1995
530.1'1—dc20   95-38236
                                                                     CIP

Manufactured in the United States by LSC Communications
68895X03   2017
www.doverpublications.com

Preface

THIS BOOK is intended to fill a need for an elementary textbook on special relativity.

The original material was developed for a junior course in mechanics for physics majors. It was revised for use by seven entering freshmen in a seminar taught at the Massachusetts Institute of Technology under the auspices of the Science Teaching Center. The final form of the notes was developed for a one semester-hour course at the University of Illinois in special relativity, for which the prerequisites are one semester of freshman mechanics and concurrent registration in later semesters of elementary physics. Each chapter corresponds roughly to one lecture in that course. While conceived as an elementary text, it is hoped that this book will introduce relativity to any beginner in the field, although readers with more advanced preparation will find the mathematics rudimentary and the examples numerous.

The treatment is considerably more lengthy than standard treatments at this level. It is hoped that the book thereby becomes adequate for self-study. The only mathematics necessary is algebra and little attempt has been made to introduce higher mathematics in order to simplify the formalism. The first concepts are introduced from the "gedanken experiment" approach, but later in the book energy is introduced more formally as a logical consequence of conservation of momentum and the symmetry of moving frames of reference.

The historical introduction in the first chapters is made with an eye to subsequent use in the book and with no idea that it gives a balanced historical approach. For this, the reader is referred to *Einstein's Theory of Relativity* by Max Born (Dover Publications, New York, 1962). History has been distorted toward an overemphasis on the Michelson-Morley experiment and mechanics, and away from electrodynamics, on the grounds that elementary students have little familiarity with or faith in their ability to manipulate fields. There has been no attempt to discuss the philosophical implications of relativity nor is this book in any sense a "popular" treatment. For this, the reader is referred to Einstein's *Relativity; the Special and the General Theory, a Popular Exposition* (15th ed., Methuen, London, 1954). The emphasis is on a real working knowledge at an elementary level. To this end the treatment includes three-dimensional space rather than motion in one direction only, since most real problems seem to occur in three dimensions!

When Einstein wanted a familiar example of a "rapidly moving object" he chose the railroad train. Times have changed. The reader will find rocketships used here. But please be reminded that rockets still go too slowly for relativistic considerations to enter in their navigation. Therefore, when rocketships are used in exercises and examples, it is simply as a more modern version of a "rapidly moving object." In order to find examples which are not too pedantic, liberal use has been made of elementary particle physics. It is elementary particle physicists who find special relativity a part of their daily lives. If you are a reader who finds a muon or K^+ meson a very new thing, do not panic. You need know almost nothing about such particles. For further information than that given in this book, there are several paperbacks on the elementary particles: *Tracking Down Particles* by R. D. Hill (W. A. Benjamin, Inc., New York, 1963), *Elementary Particles* by D. H. Frisch and A. M. Thorndike (Van Nostrand, Princeton, New Jersey, 1964), and *The World of Elementary Particles* by K. W. Ford (Blaisdell, New York, 1963).

A word of caution is in order concerning the use in this book of the term *proper time*. The proper time interval between two events is here defined to be the interval between them as measured by a single clock which is present at both events; the clock has moved

Preface

from one to the other; it is the time for a race as measured by the runner. Throughout the treatment it is tacitly assumed that the clock moves between the events at a constant velocity. Therefore the proper time interval between two events becomes a unique property of the events. Our use of proper time is thus a special case of the more usual definition of the term that allows accelerated motion and that is a property not only of the events, but also of the world line of the clock joining them. It is easy to generalize this restricted use to infinitesimal intervals and thence to the conventional proper time. I have found this treatment a good one for teaching, and I hope the simplification is permissible on this basis.

I should like to thank C. Sherwin for introducing me to the pedagogical use of proper time and G. Ascoli for the basic idea behind the Twin Paradox treatment. Especially I commend the patience of the many students who have suffered through the development of these notes; to them, much thanks.

JAMES H. SMITH

Urbana, Illinois
June 1965

Contents

Preface v

INTRODUCTION   *1*

1. CLASSICAL RELATIVITY AND THE RELATIVITY POSTULATE   3

   1-1. The Postulate of Relativity   3
   1-2. Frames of Reference   5
   1-3. Inertial Frames of Reference   8
   1-4. The Conservation of Momentum in Different Frames of Reference   9
   1-5. Absolute Velocity vs. Relative Velocity   12
   1-6. The Relative Speed of Two Observers as Measured by Each of Them   13
   Exercises   15

2. LIGHT WAVES AND THE SECOND POSTULATE   *19*

   2-1. Waves   19
   2-2. Periodic Waves: Frequency and Wavelength   21
   2-3. The Doppler Effect   22
   2-4. Light is a Wave   26
   2-5. Aberration of Light   27
   2-6. Motion Through Space   29

2-7. The Michelson-Morley Experiment	34
2-8. The Second Postulate	41
2-9. Experimental Proof of the Second Postulate	42
2-10. A Corollary to the Second Postulate	46
Exercises	46

3. TIME DILATION: PROPER AND IMPROPER TIME — 49

3-1. Gedanken Experiments	49
3-2. Measurement of Proper and Improper Time Intervals	50
3-3. Do Moving Clocks Run Slow?	54
3-4. Is the Clock Mechanism Affected by Motion?	56
3-5. Real Experiments in Time Dilation	57
Exercises	60

4. LENGTH MEASUREMENTS — 63

4-1. Length Contraction	63
4-2. The Second Arm of the Michelson-Morley Apparatus	64
4-3. A Third Length Measurement	66
4-4. The Length Paradox and Simultaneity	68
4-5. Lengths Perpendicular to Their Motion Do Not Change	73
4-6. A Summary	74
Exercises	74

5. VELOCITY AND ACCELERATION — 77

5-1. Adding Velocities	77
5-2. Emission of Light by Moving Objects	81
5-3. Acceleration	83
5-4. The Ultimate Speed	87
5-5. The Doppler Effect	88
Exercises	90

6. THE TWIN PARADOX — 93

6-1. Statement of the Paradox	93
6-2. The Solution in Terms of Time Dilation	95
6-3. The Solution in Terms of "Heartbeats"	97
6-4. Solution in Terms of "Heartbeats" Counted by the Outbound Pilot	98
6-5. Discussion and Experiment	100
Exercises	102

7. THE LORENTZ TRANSFORMATION AND NOTATION — 103

7-1. The Lorentz Transformation	103

Contents

7-2. A Matter of Notation	104
7-3. Use of the Lorentz Transformation	106
Exercises	109

8. PROPER- OR FOUR-VELOCITY — 113

8-1. Two Kinds of Velocity	113
8-2. Addition of Velocities Formula for Proper Velocity	115
8-3. An Example	121
Exercises	122

9. MOMENTUM AND ENERGY — 124

9-1. Non-Relativistic Conservation of Momentum	124
9-2. How do we Choose a Relativistic Expression for Momentum?	126
9-3. Relativistic Momentum Conservation	130
9-4. A New Conservation Law—Energy	132
9-5. An Example	135
9-6. Momentum and Energy: A Summary	139
9-7. Relativistic Momentum and Energy—Experiment	139
9-8. An Example: A Symmetrical, Elastic Collision between Equal Mass Particles	144
9-9. An Example: A Head-on Elastic Collision	145
9-10. Energy and Momentum in Two Different Frames of Reference	148
Exercises	148

10. PARTICLES OF ZERO MASS — 151

10-1. Light Flashes as "Particles"	151
10-2. Photons	152
10-3. Other Particles of Zero Mass	153
10-4. An Example: An Atom Absorbs Light	154
10-5. An Example: Decay of the K^+ Meson	155
Exercises	157

11. CENTER-OF-MASS AND PARTICLE SYSTEMS — 159

11-1. When is an Object at Rest?	159
11-2. Total Momentum and Energy of a Group of Particles	160
11-3. The Center-of-Mass Frame of Reference	161
11-4. Importance of Energy in the Center-of-Mass	163
11-5. An Example: A Collision between Equal Mass Particles	164
Exercises	171

12. FOUR-VECTORS — 177

12-1. The Energy-Momentum Four-Vector	177
12-2. The Lorentz Transformation as a Rotation in Four Dimensions	178
12-3. Ordinary Scalars	180

12-4. Four-Scalars or Lorentz Invariants	182
12-5. Four-Vectors	183
12-6. An Example: Elastic Scattering of Equal Mass Particles	185
Exercises	188

13. ELECTRIC AND MAGNETIC FIELDS AND FORCES   189

13-1. Electric and Magnetic Fields between Plane Sheets of Charge	189
13-2. Moving Condensers	191
13-3. The Field of a Moving Point Charge	192
13-4. Gauss's Law for a Moving Charge	195
13-5. Electrical Forces	199
13-6. Relativistic Force	201
13-7. Centripetal Force and Magnetic Deflection	202
Exercises	205

Appendix A. APPROXIMATE CALCULATIONS IN RELATIVITY   *207*

Appendix B. A SUMMARY OF RELATIVISTIC FORMULAS   *209*

Appendix C. A TABLE OF PARTICLES   *212*

Index   215

Introduction

BY THE YEAR 1900 PHYSICISTS HAD HAD experience with Newtonian mechanics for over two centuries. Its predictions were not only satisfyingly simple, but they were borne out by extensive terrestrial experiments as well as minutely detailed astronomical observations. The other great fields of physical research were electricity, magnetism, optics, and thermodynamics. In the latter part of the nineteenth century thermal phenomena had been shown to be susceptible of a mechanical interpretation, and the genius of Maxwell had shown that light was an electromagnetic wave and therefore that optics, electricity, and magnetism were really different aspects of one unified electromagnetic theory. It is probably a pardonably small simplification, then, to say that at the beginning of the present century, all physics was encompassed in the two great theories of mechanics and electromagnetism.

But there was trouble. These two theories as they stood were basically inconsistent. Not many recognized it, but the trouble was there, and the seeds of relativity were sown. No one can say, of course, how long it would have taken for relativity to be developed without the ideas of Einstein, but there was other contemporary, independent work, notably that of Lorentz and Poincaré, which contained many of the ideas which are now known to be consequences of the theory of relativity. It took the keen insight of Einstein, however, to see that the troubles were more basic than those of either mechanics or electromagnetic theory, and were concerned with our most elementary ideas of space and time. He reduced the apparent inconsistencies between mechanics and electromagnetism to two postulates. The first was so fundamental to mechanics and the second to electromagnetism, that they seemed necessary to the success of the two highly successful theories. The demand that the two postulates be consistent led to a new formulation of ideas about space and time, the special theory of relativity. The two postulates are:

1. The results of all experiments performed entirely within a certain frame of reference are independent of any uniform translational motion of that frame of reference.
2. In any frame of reference the speed of light is independent of the speed of its source.

The first postulate is usually called the relativity postulate. The first chapter will discuss the meaning of this postulate and how it stems from our everyday ideas of classical mechanics. The second postulate is often called the postulate of the constancy of the speed of light. The second chapter of the book will show how this postulate, strange as its results may be, comes from our commonly accepted ideas of the behavior of waves, coupled with necessary experimental information on the nature of light waves. The remainder of the book will be devoted to exploring the consequences of these postulates for our ideas about space and time, as well as their consequences for a proper formulation of mechanics.

1

Classical Relativity and the Relativity Postulate

1-1 THE POSTULATE OF RELATIVITY

LIFE ON A LARGE OCEAN LINER goes on very much as it does on shore. People swim in the pool, play shuffleboard on deck, eat in the dining room. Even on a jet airplane going at 600 mi/hr it takes no extra effort to eat dinner. Coffee pours just as it does for someone at rest on the earth. The point is that, although we think of these vehicles as being in rapid motion, it seems to make no difference in the behavior of commonplace objects when such behavior is referred to the moving system—we sometimes say: when such behavior is viewed from the *moving frame of reference*. We are not surprised at this; we expect it. In fact, if we stop to think a moment, we see that any other situation would be very peculiar indeed. For the speeds of these vehicles we have just mentioned are trivial compared to the speed of the earth about the sun or the sun through the galaxy or the galaxy . . . . In fact, we soon run into difficulty. We do not know how to measure a velocity at all unless we refer it to something, i.e., to some frame of reference. It would be very peculiar if the mechanical behavior of objects depended on the speed with which we were moving, i.e., if the laws of physics were different for observers moving with different speeds. It would mean, for instance, that a billiard player would have to adjust his style of play to the season of the year, because the motion of the earth about the sun requires that its velocity differ by 60 km/sec at 6-month intervals. It is our common experience that no such adjustment is necessary, and that no experiment, however precise, has shown any such difference at all. That is all the postulate of relativity means.

It does, of course, go somewhat beyond our ordinary experience because it says that *all* experiments will give results independent of the velocity of the frame of reference in which the measurements are made. All of our experience is limited—and that includes laboratory experience in physics. All that experience can say is that nobody has yet performed an experiment that contradicts this postulate of relativity; tomorrow someone may do so. We will then have to revise our opinions. But for now we take the postulate at its face value, without reservation, and admit that no experiment can be performed which will give different results when performed in two laboratories moving uniformly with respect to one another.

The word "experiment" is used in the first postulate in a somewhat restricted sense. To pursue the analogy of billiards it means the making of a particular shot in a particular way. The "results" of the "experiment" are simply the consequences of the shot. The experiment consists of setup *and* results, both related to a particular frame of reference. The first postulate says that if the setup is made in the same way, the results will be the same whether the billiard table is "fixed" on the earth or carried in a speeding plane. On the other hand, measuring the "speed of the earth" is not an experiment in this sense. There is no particular set of initial conditions; there are no consequences to be determined. Clearly, the speed of the earth depends on the particular frame of reference in which the measurement is made.

1-2 FRAMES OF REFERENCE

In the last section we used the term *frame of reference*, and we implied the measurement of motion with respect to a frame of reference. We will be using this term frequently, and in this section we will formulate our ideas somewhat more precisely.

Although the term is usually applied to the entire situation in which a particular experiment is performed, it is probably helpful to think of a frame of reference as the *coordinate system* with respect to which measurements are made. To say that an automobile is moving at 60 mi/hr in the frame of reference of the earth, or simply with respect to the earth, implies that the automobile passed one point fixed on the earth at one instant and passed another point fixed on the earth 60 mi distant from the first, one hour later.[1] When we say that a man walks down the aisle of a jet airliner at 3 ft/sec, we imply that the measurement was made in the frame of reference of the airliner, i.e., with respect to a coordinate system fixed to the airliner. With respect to a coordinate system fixed to the ground, the man is possibly moving 900 ft/sec. We see, then, that even in this very simple situation, the description of the motion of an object (the man) depends on the frame of reference from which it is viewed (the plane or the earth).

Suppose that there are two coordinate systems moving with respect to one another with a speed v. We call one the O system and the other the O' system. The whole O' system is moving to the right with a speed v along the positive x-axis of the O system. Conversely, the O system is moving toward negative x' as measured in the O' system. Figure 1-1 shows the system at several different times. Figures 1-1a, b, and c have been drawn as if the O system remained fixed on the page and Figures 1-1d, e, and f as if the O' system remained fixed, but it must be *clearly* understood that the sequence depicted by Figures 1-1a, b, and c is the *same* as that depicted by Figures 1-1d, e, and f. The only thing with physical content is that

[1] For the moment, we may be careless about the clock, although when we think relativistically, we will also imply that the frame of reference of the earth also means clocks fixed to the earth.

O and O' are moving *apart* with the speed v. Figure 1-1c looks just exactly like Figure 1-1f.

Now let us suppose that some object A starts in the O system and moves from the origin ($x = 0$) at the time $t = 0$ and later is found to be at the point (x,y) at the time t. This is shown in Figures 1-2a and 1-2b. Furthermore, suppose the origin of the coordinate system O' happens to coincide with the origin of O at $t = 0$ but is moving along the x-axis with the speed v so that by the time t it has advanced a distance vt. This is shown in Figures 1-3a and 1-3b. It would be equally correct to say that O had moved a distance vt toward negative x' measured with respect to O'. The object A is found a distance

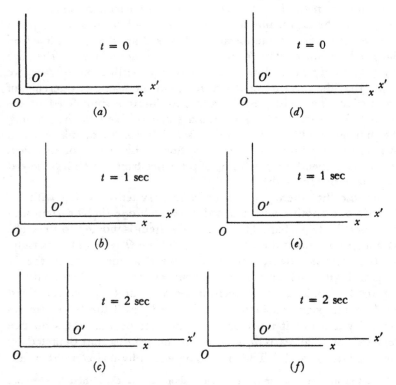

Figure 1-1. The O and O' frames of reference are moving apart. In (a), (b), and (c) this motion has been drawn as if the O frame remained fixed; in (d), (e), and (f) the same relative motion has been shown as if the O' frame remained fixed.

The Relativity Postulate

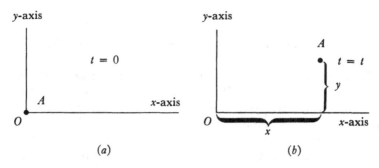

Figure 1-2

x' from the O' origin at the time t, and a glance at Figure 1-3b shows clearly that $x = x' + vt$. Since the relative motion of the coordinate systems occurred along the x-axis, it is equally clear that $y = y'$, and if a third coordinate were shown, $z = z'$. Although it is, for present purposes, meaningless to distinguish them, we will explicitly state that clocks in both coordinate systems read alike and therefore $t = t'$. Summarizing, we have

$$x' = x - vt$$
$$y' = y$$
$$z' = z$$
$$t' = t \tag{1-1}$$

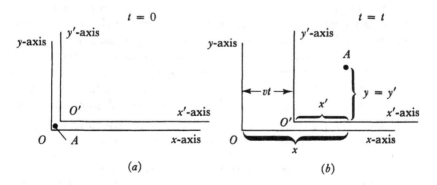

Figure 1-3

1-3 INERTIAL FRAMES OF REFERENCE

In what we have said so far it seems that any frame of reference is equivalent to any other. That is certainly not so. Coffee may pour in a smoothly riding jet plane just like coffee on the ground, but if the ride is bumpy, allowances must be made. No fancy apparatus is necessary. One's stomach is an excellent indicator. The difference between a smooth ride and a bumpy one is clearly one of *acceleration*. All experiments, so reads the first postulate, give the same results in uniformly moving coordinate systems. It follows that no experiment will detect uniform motion. Even one's stomach detects accelerated motion.

For instance, take Netwon's First Law, the Law of Inertia: "A body at rest will remain at rest, or a body in uniform motion will remain in uniform motion unless acted on by a force." When that motion is measured with respect to a bumpy jet airliner, the law simply is not true. It is not true on a merry-go-round. Put a marble on the floor of a merry-go-round and it will not remain at rest with respect to the merry-go-round. It will immediately accelerate toward the outside. What does it mean, then, to say that Newton's First Law is *true*? When we say that it is true, we simply mean that there *are* coordinate systems where it is true. For most purposes the earth is such a coordinate system.[2] Sensitive measurements can, nevertheless, detect accelerations due to its rotation. Astronomical observations lead us to believe that the law is more nearly true when referred to the coordinate system of the fixed stars. This discussion could lead us far afield. We will simply assume that there exist coordinate systems where Newtonian mechanics works. We call such frames of reference *inertial frames of reference* or sometimes just *inertial frames*. In such frames of reference certain laws of physics hold. What our experience tells us, and the relativity postulate

[2] Actually, a marble released near the surface of the earth doesn't remain at rest; it falls. One can take the point of view—the one assumed in this book—that the frame of reference of the earth is an inertial frame, but that an explicitly recognized force, gravity, is acting on all objects near the earth's surface. On the other hand, one can take the point of view of general relativity that a stationary frame of reference near the earth's surface is not an inertial frame.

The Relativity Postulate

states explicitly, is that in all frames of reference moving uniformly with respect to an inertial frame, the same laws of physics hold. In this book we shall be almost entirely concerned with such inertial frames. How the laws of physics must be formulated in noninertial, i.e., accelerated, systems is the province of the more complex general theory of relativity.

1-4 THE CONSERVATION OF MOMENTUM IN DIFFERENT FRAMES OF REFERENCE

Our discussion of experiments in different frames of reference has, up to now, been rather general and qualitative. We have said that the relativity postulate arose largely from important ideas in classical mechanics, and it is the purpose of this section to explore this connection in more detail. We are going to use, as a simple example, an experiment in conservation of momentum performed in a laboratory fixed on the earth and in a second laboratory moving uniformly past it at a speed v in the positive x-direction with respect to the earth. This second "laboratory" might be a speeding train.

We will assume that the laboratory fixed on the earth is an inertial frame of reference, i.e., that the familiar laws of mechanics hold there. Among these is the law of conservation of momentum. We will show that if momentum is conserved on the earth, it is also conserved on the train.

Suppose that an observer on the train, Figure 1-4a and b, performs a simple collision experiment between objects of masses m_1 and m_2. For simplicity, suppose all velocities are in the x-direction. The object m_1 has a velocity u_1' before the collision and U_1' after it; m_2 has velocities u_2' and U_2'. The primes indicate that these velocities are to be measured with respect to the train. Since we claim to be in doubt about the validity of conservation of momentum on the train, we do not claim to know the relations between the velocities before and after the collision directly. We do, however, know that momentum is conserved in the frame of reference of the ground. We therefore attempt to describe the same experiment in that frame of reference.

To an observer on the ground watching this experiment, Figure

10   Introduction to Special Relativity

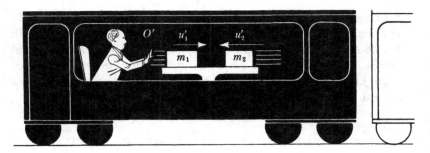

(a)

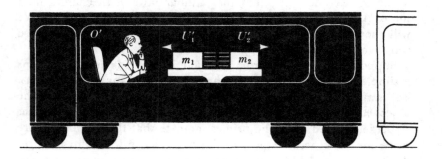

(b)

Figure 1-4. (a) An observer O' riding on a train performs a collision experiment between two objects m_1 and m_2 moving at velocities u_1' and u_2' with respect to the train. (b) After the collision, the two objects bounce apart with velocities U_1' and U_2', still measured with respect to the train. (c) Meanwhile, to an observer O standing beside the track, the same collision appears to be one between a mass m_1 with velocity $u_1 = u_1' + v$ and a mass m_2

The Relativity Postulate

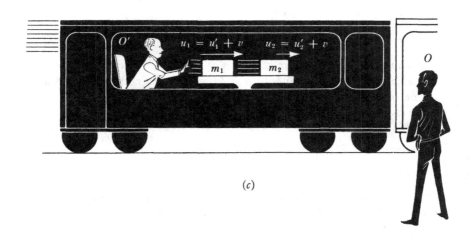

(c)

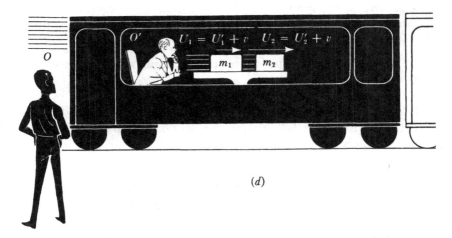

(d)

with a velocity $u_2 = u_2' + v$ where v is the velocity of the train past O. (d) After the collision, the masses move with velocities $U_1 = U_1' + v$ and $U_2 = U_2' + v$ with respect to O. Since O is supposed to be in an inertial system where momentum is conserved, $m_1 u_1 + m_2 u_2 = m_1 U_1 + m_2 U_2$. It follows that $m_1 u_1' + m_2 u_2' = m_1 U_1' + m_2 U_2'$, i.e., that momentum is conserved on the train.

1-4c and d, the object m_1 has a velocity

$$u_1 = u_1' + v \tag{1-2}$$

Equation 1-2 is a familiar one, but is proved in Exercise 1 from Equation 1-1. Since momentum is known to be conserved in the frame of reference of the ground

$$m_1 u_1 + m_2 u_2 = m_1 U_1 + m_2 U_2 \tag{1-3}$$

Substitution of Equation 1-2 into Equation 1-3 gives

$$m_1 u_1' + m_1 v + m_2 u_2' + m_2 v = m_1 U_1' + m_1 v + m_2 U_2' + m_2 v$$

where equations similar to Equation 1-2 have been used for the other velocities. Canceling the terms containing v,

$$m_1 u_1' + m_2 u_2' = m_1 U_1' + m_2 U_2' \tag{1-4}$$

This last equation expresses conservation of momentum in the frame of reference of the train.

Proving that momentum was conserved on the train if it was conserved on the ground, i.e., if momentum was conserved in one frame of reference then in all frames moving uniformly with respect to it, depended primarily on two things:

1. The form of the dependence of momentum on velocity, namely momentum $= mu$.
2. The relation between the velocities of an object as seen by two different observers, i.e., $u = u' + v$ which, in turn, depends ultimately on Equation 1-1 relating the positions of an object as seen by two observers.

1-5 ABSOLUTE VELOCITY VS. RELATIVE VELOCITY

In the last section we saw that an experimenter on a train would find that momentum was conserved even though the train was moving rapidly. In fact, we said it could be shown that *any* mechanics experiment he could perform would give the same result as it would if performed on the earth. The importance of this idea for mechanics

The Relativity Postulate

was well understood. Einstein recognized it as a general property of nature applying to all types of experiments, not merely those of mechanics.

Now if the results of *any* experiment performed on the uniformly moving train will give the same results as an identical experiment performed on the earth, then it is clear that there is no way for the experimenter on the train to tell whether he is moving—at least by the performance of an experiment done entirely on board the train. This leads to an alternative statement of the relativity postulate which is often convenient to remember and which is somewhat closer to the popular conception of relativity:

No experiment can be performed which will detect an absolute velocity through space.

Of course, the simplest experiment in the world, looking out the window, will detect relative motion between the train and the earth, but that is all. The man riding on the train is unable to tell whether he is moving or the earth is moving past him. Even the jiggles and joggles are not enough to assure us that the train is moving because they would still be there if, in some cosmic joke, a giant were pulling the "rug" out from beneath the train!

1-6 THE RELATIVE SPEED OF TWO OBSERVERS AS MEASURED BY EACH OF THEM

There is a corollary to the first postulate which is probably so self-evident that it needs no proof, and yet it is so important for some of our future considerations that this section is devoted to a discussion of it. It is this:

If the uniform speed of an observer O' relative to an observer O is v, then the speed of O relative to O' is also v.

This only means that if a man standing beside the highway sees a car going down the highway at 60 mi/hr, the driver of the car sees the first man passing at 60 mi/hr.

As a further example, suppose two trains are passing each other, and a man on train A has a device for measuring the speed of train B past train A. (See Figure 1-5.) He finds v_A (i.e., v_A is the *reading* of the device on train A). He has arranged for an identical device to be placed aboard train B and for the information to be telemetered back to him. The result of this experiment is that the remote device reports v_B (the *reading* of the device on train B).

Now think about this whole operation from a different point of view. A second man stands beside the device on train B, which is reading v_B, and receives a report from the first device on train A,

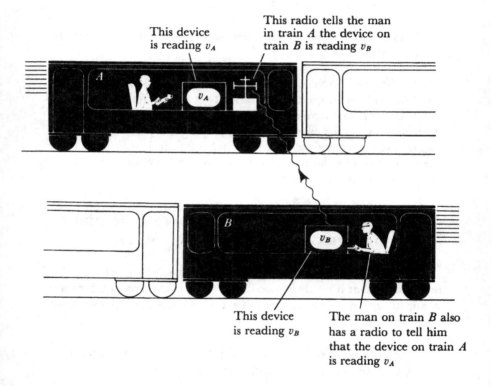

Figure 1-5. *A man on train A and another on train B are doing identical experiments. The one on train A says:* "When my device reads v_A for the relative velocity between the trains, the identical device aboard the other train reports a relative velocity v_B." *The man on train B says:* "When my device reads v_B, the other device reads v_A."

The Relativity Postulate

which is reading v_A. Let us suppose that the readings are not identical even though the devices are functioning properly. In this case one reading must be greater than the other, say $v_A > v_B$. Therefore the man on train A observes: "The remote device reports a lesser reading than the device at rest beside me." The man on train B observes: "The remote device reports a greater reading than the device at rest beside me." Yet the two observers are doing the same experiment and must, on the basis of the relativity postulate alone, get the same answer. Our supposition that $v_A > v_B$ is therefore incorrect and there is only one [3] self-consistent possibility, namely, that the readings are the same.

It should be emphasized that this discussion has been made on the basis of the relativity postulate alone, and does not depend on the form of Equations 1-1 or 1-2. It is, therefore, a statement that will remain true in special relativity when other things, which many students would consider even more self-evident, do not.

Exercises

1. Equation 1-1 gives the *position* of an object in the O' frame in terms of its position in the O frame as a function of time.

a. Assume that the object is moving in the O' frame with a uniform velocity u' and thus reaches x_1' at t_1' and x_2' at t_2' where $u' = (x_2' - x_1')/(t_2' - t_1')$. Find the positions of the object in O at the times $t_2 = t_2'$ and $t_1 = t_1'$, and from this the velocity u of the object in O, i.e., verify Equation 1-2.

b. Repeat *a* simply by differentiating Equation 1-1.

2. Figure 1-6 shows a diagram of the successive positions of two objects colliding and bouncing apart. The figure can be considered a photograph taken by repetitive flashes of light spaced at $\frac{1}{10}$ sec intervals. Figures 1-6a and 1-6b show the same collision, but one was taken with a camera which was moving uniformly. The scale of distance is shown in the figure. The object marked with a small circle has a mass of 1 kg.

a. Using Figure 1-6a and the conservation of momentum, determine the mass of the object marked with the cross.

b. Show by direct measurement, and using the mass determined in *a*, that momentum is also conserved when measurements are made in the frame of reference of Figure 1-6b.

---

[3] See Exercise 6.

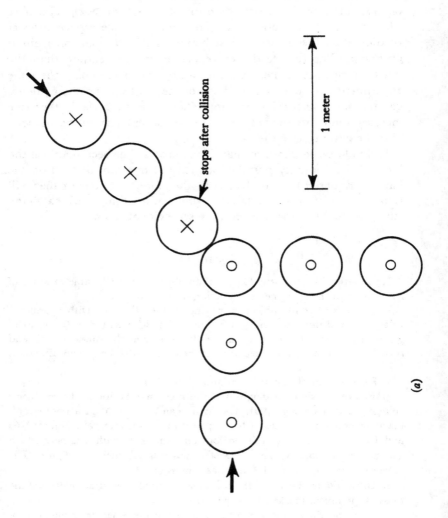

(a)

Figure 1-6

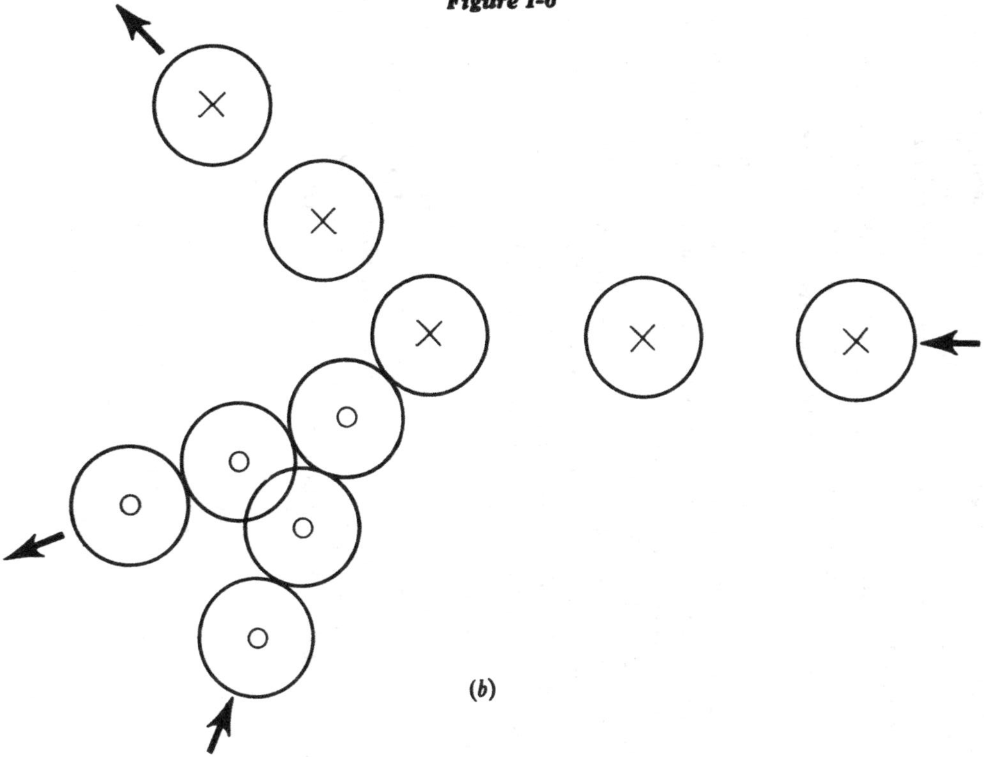

(b)

c. Determine the magnitude and direction of the velocity of the camera used to take the picture 1-6*b* as measured in the frame of reference of 1-6*a*.

d. Which picture, 1-6*a* or 1-6*b*, was made with the moving camera?

3. Again referring to Figure 1-6, how much kinetic energy has been converted into thermal energy in the collision of 1-6*a*? in the collision of 1-6*b*? Could the temperature rise of the objects be used to determine which frame of reference was moving?

4. Show analytically, using Equation 1-2, that the amount of thermal energy produced in a collision is independent of the frame of reference from which the collision is viewed. Did you have to assume the conservation of momentum?

5. Suppose an object has an acceleration a when viewed in a frame O. Use Equation 1-1 to find its acceleration as viewed in a frame O' moving uniformly with respect to O. What does this say about the force on the object as seen by an observer in O and one in O'?

6. In Section 1-6 a device on train B reported back a relative speed reading v_B when a device on train A read v_A. Actually all we know is that for every v_A there will be some v_B, i.e., v_B is a function of v_A. Similarly, from the point of view of the other observer, v_A is a function of v_B. The relativity postulate says that the *functions* must be the same, i.e.,

if $v_B = F(v_A)$ then $v_A = F(v_B)$

Clearly one solution of these equations is that the function is just the variable itself: $v_A = v_B$. Another solution that differs only trivially because it amounts to a choice of which direction each observer calls positive is: $v_A = -v_B$. There is still another solution; can you find it? On what basis do we reject this other solution as a physical possibility?

2

Light Waves and the Second Postulate

IN CHAPTER 1 WE SAW that a very important property of mechanics is that its laws are identical for observers moving uniformly with respect to one another. Although this property of mechanics had been known for years, until the time of Einstein's formulation of the first postulate, it was not at all obvious that the same property should be a part of physics in general. In fact, it was believed that a carefully performed experiment in optics would, indeed, determine the velocity of the earth through space, and the failure of such experiments was one of the main paths which ultimately led to the special theory of relativity. The subject of this chapter is the review of those properties of waves necessary to an understanding of these experiments and how they led to the postulates of relativity.

2-1 WAVES

Everyone has, at one time or another, dropped a pebble into smooth water and watched the circular rings spread out away from the source of the original disturbance. This is a typical wave motion. There is a medium, the water, which normally lies in an undisturbed, equilibrium state. When something disturbs the medium, that portion of the medium originally affected disturbs that next to it, and that, the medium next to it, and so on. The result is that a disturbance—the wave—travels out through the medium. It is easy to see that the wave carries energy with it. The pebble transmits some of its energy to the water, and some time later a leaf bobs up and down as the wave passes. The wave has carried energy from the pebble to the leaf. That "some time later" is important. It indicates what everyone has observed, namely, that it takes time for the wave to go from place to place. The wave has a speed. Since the wave is transmitted from one bit of the medium to the next, the speed with which the wave travels can depend only on the properties of the medium. Therefore, once the wave has left the immediate environment of the original disturbance, its speed cannot depend on the nature of that disturbance; it cannot, for example, depend on the speed of the source through the medium.

Some types of waves, of which water waves are an example, have speeds that depend on the particular shape of the wave. Since the shape of the wave depends on the nature of the disturbance which caused it, the speed of the wave also might be said to depend on that disturbance. Such an indirect dependence need not worry us, especially since the speed of light waves through space, which is our ultimate concern, does not depend upon wave shape.

Let us emphasize again that the medium does not move along with the wave. No particle of water originally disturbed by the pebble eventually reaches shore with the wave. Any motion of the medium is only a local one as the wave passes. It is the disturbance or, if you prefer, the energy in the wave, which moves.

There are many familiar examples of waves besides water waves. Sound is a wave in which the molecules of the air are disturbed by a source, a loudspeaker perhaps, and transmit this disturbance to their neighbors until at last the molecules next to our ears disturb our eardrums and we hear sound. A rope, or better, a spring, can be stretched out and wiggled so that a wave passes along it. Even a long relay race might be considered a wave, since no single runner finishes the whole course, but *something* goes from start to finish.

Light Waves and the Second Postulate

In all these cases we recognize the essentials of a wave: a disturbance moves from place to place with a speed that is determined by the properties of the medium itself.

2-2 PERIODIC WAVES: FREQUENCY AND WAVELENGTH

Although a wave can consist of a single disturbance which passes and then is gone, very frequently we are concerned with phenomena in which the disturbance repeats itself periodically and sends out regularly spaced waves. Figure 2-1 shows a stationary wave source S producing circular waves that travel out through the medium with a speed c. If the source produces f of these regularly spaced disturbances per second, it is said to emit a frequency f. The frequency of a sound wave determines its pitch. Middle C corresponds to 256 vibrations per second; upper C to 512; etc.

If the source S produces f waves per second, the time between the production of two successive wave crests is $1/f$. During this time the waves advance a distance λ, called the wavelength. It follows im-

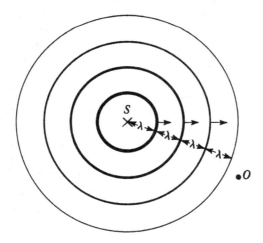

Figure 2-1

mediately that the speed is

(speed) = (distance) ÷ (time) = (frequency) × (wavelength)

$$c = \lambda \Big/ \left(\frac{1}{f}\right) = f\lambda \qquad (2\text{-}1)$$

This is a relation which is generally true of all regular periodic waves.

Now consider an observer standing at O. As the waves pass him, they disturb his eardrum if these are sound waves, or rock his boat up and down if we are considering water waves. It is clear that he will hear, or be rocked, at the same frequency as the source; but let us see how this comes about in detail. In a moment the outermost wave crest reaches him. Then, a time Δt later, the next wave crest will reach him. Since it had to travel a distance λ at a speed c, $\Delta t = \lambda/c$. The number of pulses he receives per second, i.e., the frequency he observes, is thus c/λ. Comparing this to Equation 2-1 we see that this is the frequency emitted by the source—as we suspected.

2-3 THE DOPPLER EFFECT

The observer of a wave does not always observe the same frequency as that emitted by the source. If either the source or observer is moving, he observes a different frequency. This effect—the Doppler effect—is responsible for the sudden change in pitch when an automobile sounding its horn, or a locomotive sounding its whistle, passes us.

Consider the situation shown in Figure 2-2. Here the source S is moving through the medium to the right. We assume that the medium is at rest in the figure, i.e., we are considering the problem in the frame of reference of the medium. We consider the observer O to be at rest also. The outermost disturbance shown was produced when the source was at A. Since the propagation of the wave is with respect to the medium, this outermost wave is a circle about A. The distance the wave has traveled and that it is a circle have nothing to do with the fact that the source has moved in the meantime. The other disturbances are produced at later times and are, consequently, circles centered on points farther and farther to the right. To the right of S the successive wave crests are crowded closer together,

Light Waves and the Second Postulate

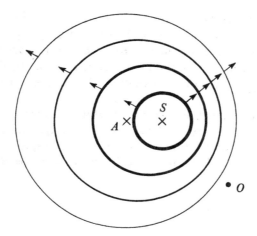

Figure 2-2

and to the left spaced farther apart, than in Figure 2-1. Since the speed of the waves has nothing to do with the velocity of the source, the waves are still going with a speed c. The frequency seen by an observer somewhere to the right will therefore be higher than that of the source. Let us see how much higher.

In Figure 2-3 we see that a wave crest 1, produced when the source was at S_1, has just arrived at an observer O. The source is moving to the right with a velocity v and the angle between source

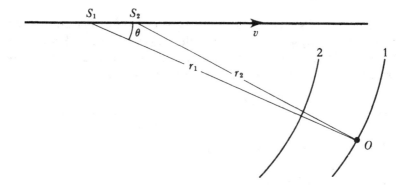

Figure 2-3

velocity and direction to the observer is θ. By the time the source produced the next wave crest 2, which has not yet reached the observer, the source had moved to S_2. Since the source is emitting waves at a frequency f_S, the time between the production of successive crests is $1/f_S$. During this time the source moves a distance v/f_S. Thus $\overline{S_1S_2} = v/f_S$.

Suppose the first wave crest was produced at time $t = 0$. It arrives at O at a time $t_1 = r_1/c$. The second wave crest was produced at a time $1/f_S$ and takes a time r_2/c to get to O, thus, arriving at a time $t_2 = 1/f_S + r_2/c$. The time *interval* between arrivals of the crests at O is

$$\Delta t = t_2 - t_1 = \frac{1}{f_S} + \frac{r_2}{c} - \frac{r_1}{c}$$

From the law of cosines we have

$$r_2 = \sqrt{r_1^2 + \overline{S_1S_2}^2 - 2r_1\overline{S_1S_2}\cos\theta}$$

This exact expression, it turns out, leads to a fairly complicated and unilluminating expression. In most practical cases, O is much farther from the source than the distance the source moves between production of successive waves—in fact, if this is not so, it makes little sense to talk about a single frequency observed by O. We have then, $\overline{S_1S_2} \ll r_1$ and (see Appendix A)

$$r_2 \approx r_1 - \overline{S_1S_2}\cos\theta$$

Another way of seeing the same thing is to look at Figure 2-3 and realize that if $\overline{S_1S_2} \ll r_1$, the triangle S_1S_2O is a very skinny one. Projecting $\overline{S_1S_2}$ onto S_1O shows immediately that r_1 is very nearly composed of $r_2 + \overline{S_1S_2}\cos\theta$. In either case, we have

$$\Delta t = \frac{1}{f_S} + \frac{r_1 - \overline{S_1S_2}\cos\theta}{c} - \frac{r_1}{c} = \frac{1}{f_S} - \frac{\overline{S_1S_2}}{c}\cos\theta$$

Substituting $\overline{S_1S_2} = v/f_S$

$$\Delta t = \frac{1}{f_S}\left(1 - \frac{v}{c}\cos\theta\right)$$

Light Waves and the Second Postulate

If Δt is the time between the arrival of two waves at O, the frequency observed at O is

$$f_O = \frac{1}{\Delta t} = \frac{f_s}{1 - \frac{v}{c}\cos\theta} \qquad (2\text{-}2)$$

If $\cos\theta$ is positive, i.e., the observer sees the source approaching, the observed frequency f_O is higher than the source frequency. As the source passes the observer, $\cos\theta$ becomes negative and the frequency is lower (just as we observe in the case of the passing locomotive).

Figure 2-4 shows the case where the observer is approaching the source at a velocity v. The reader is asked to fill in the blanks in the following sentences; the answers will be found at the end of the chapter exercises. In this case we again suppose that a wave has just reached the observer at O_1 and that this wave was emitted at time $t = 0$.

The wave arrives at O_1 at the time $t = $ _____ (a).

The next wave is emitted from S at time $t = $ _____ (b) and arrives at O_2 at $t = $ _____ (c).

The time between the observation of two successive waves by O is $\Delta t = $ _____ (d).

The law of cosines gives $r_2 = $ _____ (e).

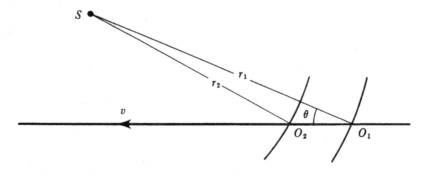

Figure 2-4

If the distance $\overline{O_1O_2}$ is very small compared to r_1, r_2 is approximately equal to _____ (f).

Then, since $\overline{O_1O_2}$ is the distance traveled by O in the time between the receipt of the two waves, $\overline{O_1O_2}$ = _____ (g).

Solving for Δt, and setting $f_O = 1/\Delta t$ we get

$$f_O = f_S \left(1 + \frac{v}{c} \cos \theta\right) \tag{2-3}$$

2-4 LIGHT IS A WAVE

All the waves which we have considered so far are basically mechanical in nature, i.e., they involve disturbances in a clearly mechanical medium. On a microscopic scale they are subject to all the laws of mechanics. They obviously, then, obey the postulate of relativity along with all other mechanical phenomena. However, there is another exceedingly important wave phenomenon, light, that is not so clearly mechanical. At one time light had been thought to consist of streams of particles that were subject to the laws of mechanics. Early in the nineteenth century, however, the evidence that light was a wave of some kind had become overwhelming, but it was not until the work of Maxwell that the nature of the wave became apparent. He showed theoretically, and many experiments later confirmed, that light was an electromagnetic wave. Maxwell showed that a changing electric field could produce a magnetic field and a changing magnetic field could produce an electric field, etc., the whole system propagating itself in the nature of a wave. This picture is a considerable oversimplification of the facts, but it contains the basic idea. Furthermore, from the results of purely electrical experiments—measuring the capacity of a well constructed condenser, for example—it was possible to predict the speed with which these waves travel,

$c = 3 \times 10^8$ meters/sec [1]

[1] It is common practice to take the speed of light to be 3×10^8 meters/sec. However, it must be remembered that this is an experimentally measured number $2.997925 \pm .000003 \times 10^8$ meters/sec.

Light Waves and the Second Postulate

This was in complete accord with the speed of light, already measured.

The question immediately arises: With respect to what should the speed of light be measured? All the waves previously considered have a characteristic speed which is the speed of the wave with respect to the medium. It was natural at that time to assume that light traveled in some medium, and this medium was called the *ether*. Light, it was assumed, traveled with a speed of 3×10^8 meters/sec with respect to the ether.

The ether had some rather remarkable properties. It seemed to fill all space. It was not, in the ordinary sense, a mechanical substance since it filled a vacuum as evidenced by the passage of light through the most perfect vacuums. It was, therefore, weightless. It extracted no energy from the light that passed through it, and was, therefore, called perfectly elastic. On the contrary, a wave passing across water sooner or later dissipates and gives up its energy to heating the water. Furthermore, it penetrated matter freely. This was shown by the famous astronomical observations of aberration. Since this particular property of the ether has more to do with our story than any other, we will study it in detail.

2-5 ABERRATION OF LIGHT

Figure 2-5a illustrates rays of light coming down from a distant star and the ray (c) entering the objective of a telescope at O and proceeding to the eyepiece E. The light goes straight down the telescope tube and the star will be seen. Suppose, however, that the telescope is moving rapidly to the right, as in Fig. 2-5b, so that by the time the light that entered at O has reached E, the telescope has moved from OE to $O'E'$. The light would have hit the side of the tube and the star would not be seen. In order to see the star, it is necessary to tilt the telescope as in Figure 2-5c. Thus, the light that hit the objective at O goes straight down as before but, now, by the time the light has reached the bottom of the tube the telescope has moved to $O'E'$ and the light hits the eyepiece and is seen. If the vertical distance the light travels through the telescope is l, it takes a time $t = l/c$ to traverse the telescope. During this time the telescope advances the distance $s = vt = vl/c$, where v is the speed of the telescope. Thus, the telescope must be tilted at an angle α where

$$\tan \alpha = \frac{v}{c} \tag{2-4}$$

The earth goes around the sun at a speed of approximately 3×10^4 meters/sec. Therefore, when pointed at stars at the most advantageous position, the telescope must be tilted at an angle of about 20 sec of arc to compensate for this effect, called the aberration of light. Aberration is easily observable and, in fact, was first reported by Bradley as early as 1729.

It should be pointed out that aberration is useless for measuring a uniform motion through space. If the earth's motion were uniform, the "apparent" positions of all stars would simply be shifted by a constant amount from their "true" positions, and no observation could tell the difference. However, the earth's motion is nearly circular and therefore the direction of the aberration reverses every six months and can be observed.

The importance of aberration in this discussion is that it shows

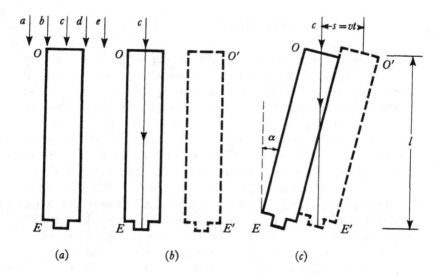

Figure 2-5

that the earth moves through the medium in which light travels. It does not drag the ether along with it, as it does its atmosphere. An illustration may clarify this. Fine raindrops fall slowly through the air; they are, in fact, dragged along with the air. If in still air the rain is falling straight, a stovepipe held vertically will allow the rain to fall straight through. If a person holds the stovepipe and walks through the rain, he must tilt it in order not to wet its sides—this might be called the aberration of raindrops. If however, a wind were blowing at the same speed the person walked, the stovepipe would have to be held vertically again. Thus, if the person created his own wind by dragging the air with him, the raindrops would fall straight down a vertical stovepipe. Indeed, if the stovepipe dragged the air inside it—as it probably does—the stovepipe should be held vertical even though moving. In all these cases there would be no aberration of raindrops.

The observation of the aberration of starlight as the earth goes around the sun thus provides information on three things:

1. It is the outstanding experiment to show that the earth really goes around the sun; and it is the Copernican rather than the Ptolemaic model of the solar system that is in accord with the facts.

2. If light travels in a medium—the ether—the ether freely penetrates telescope tubes.

3. The ether is not dragged along by the earth but, considering the results of (2), it is likely that the ether penetrates the earth and freely flows through it.[2]

2-6 MOTION THROUGH SPACE

Even though aberration turns out to be a useless tool for measuring absolute motion through the ether, it can be made immediately obvious that measurements of light can provide this information. Imagine dropping a pebble into a smoothly flowing stream. A cir-

[2] Actually the historical situation was a good bit more complicated than that presented here. Aberration experiments were performed with telescopes filled with water, and the results indicated that the supposed ether was *partly* dragged along with the water. Other experiments came to the same conclusion. Such results can be understood on the basis of non-relativistic theory, but the only completely convincing treatment comes from special relativity. Such experiments were among the many that can be said to have set the stage for relativity.

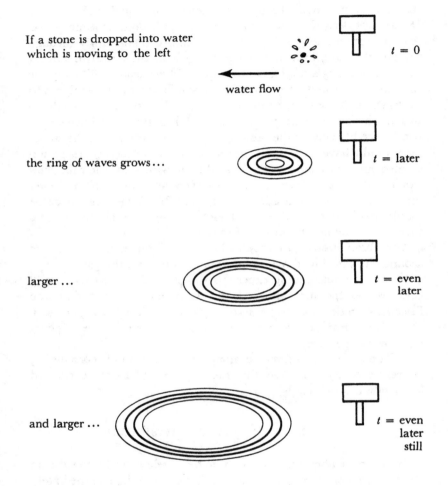

but if the speed of the water is greater than the speed of the waves with respect to the water, the wave can never progress upstream.

Figure 2-6

Light Waves and the Second Postulate

cular wave would spread out at the characteristic speed of waves in water. If the stream happens to be flowing *faster* than this, the wave cannot proceed upstream. Figure 2-6 shows such a case.

It does not matter if we consider the stream, i.e., the medium, to be moving left, or the source to the right. In other words, since we are considering the ether to be a kind of mechanical medium, the ordinary relativity of classical mechanics still holds. A look at Figure 2-6 shows that no waves propagate upstream from the source —or, in the case of light, upstream from the source of light. It is then immediately obvious that we are not moving through the ether at a speed faster than that of the speed of light through the ether. If we were, a light turned on in the middle of the room would never illuminate the "upstream" side of the room. So we see that *in principle* optical measurements can detect an absolute motion through the ether. The problem is that the speed of light is so great that the measurements are very difficult to do in practice.

Consider the schematic setup in Figure 2-7. A single light flasher F sends out a flash of light that is picked up at two detectors, D_1 and D_2. In principle, the detectors could be simply two clocks. The flash of light, when it arrived, could illuminate the clock; and a camera, continuously open, could record a picture of the two clocks, in the same way that a photographic flash lamp can be used to take a picture. Such an experiment is usually called a "gedanken" experiment, or thought experiment. It is an experiment that can be performed *in principle* but, probably, would not or could not be done that way in practice. Such "experiments" are

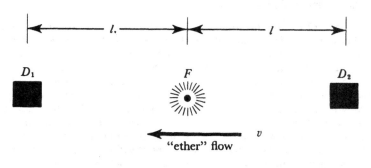

Figure 2-7

often useful in discussing physical problems and may lead to real experimental setups. We will make extensive use of such "gedanken" experiments in this book.

Let us return now to our case of Figure 2-7. Suppose the clocks are each at a distance l from the flasher. Suppose that the source is moving to the right through the ether with a speed v, or that the ether is moving to the left with speed v. The speed of light to the right would be $c - v$ and to the left $c + v$. The times for the light to get to D_1 and D_2 would be

$$t_1 = \frac{l}{c+v}; \quad t_2 = \frac{l}{c-v}$$

Therefore, the two clocks would differ in their readings when the flashes arrived by

$$t_2 - t_1 = \frac{l}{c-v} - \frac{l}{c+v} = \frac{2lv}{c^2 - v^2} \tag{2-5}$$

The most rapidly moving laboratory at our disposal is the earth itself. Since it goes around the sun at a speed of 3×10^4 meters/sec, it must pass through the ether with at least this speed at some time in the year. Therefore, suppose $v = 3 \times 10^4$ meters/sec. Take l as large as practical, say 3×10^4 meters or about 20 miles. Then

$$t_2 - t_1 = \frac{2 \times 3 \times 10^4 \times 3 \times 10^4}{(3 \times 10^8)^2 - (3 \times 10^4)^2} = 2 \times 10^{-8} \text{ sec}$$

Such a measurement could therefore show a time interval on the clocks, in principle,[3] at least.

---

[3] As we shall see, the theory of relativity predicts no time difference in any case, but perhaps it is wise to point out that even without a complete theory of relativity, electromagnetic theory would predict a zero result if the clocks were electromagnetic in nature. The problem is that even though synchronized at the central station, the clocks would gain or lose time in being carried apart in just such a way as to give a zero time delay when the experiment is performed. If, instead, the clocks were carried apart and then synchronized by sending an electric signal between them, the velocity through the ether would affect the signal so as to still maintain the zero time delay. Although this path may be used to get to the complete theory of relativity, it will not be used in this book. If, on the other hand, the clocks were mechanical, such cancellation would not be expected by pre-relativistic physics. In fact, Maxwell proposed a scheme (a

Light Waves and the Second Postulate

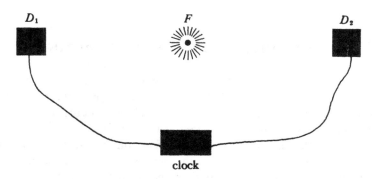

Figure 2-8

At the turn of the century, the detection of such a small time interval directly on two "clocks" would have been completely impossible. Today it would be quite easy to detect such an *interval* on a single clock; for instance, two electrical signals 2×10^{-8} sec apart are easily resolved on an oscilloscope. However, in the present example it would be necessary to set two separate clocks together at some central station and then carry them to their respective positions. During this time the clocks could not lose or gain 10^{-8} sec. To carry them apart and set them up would take about 1 hr or 3600 sec. Thus the two clocks must keep time to 10^{-8} sec in 3600, or one part in 3.6×10^{11}. Such accurate clocks are not yet available, though in a few years they may be.

But such a solution looks foolish. Could we not send out the flash, detect it with photocells at D_1 and D_2, and send electrical signals back to a central clock over carefully measured equal cables as in Figure 2-8? In such a case we would not need accurate clocks, but could detect the interval of 2×10^{-8} sec directly. The fallacy becomes apparent when we realize that the signal must come back as an electrical signal. It, too, will travel with respect to the ether, and what time the light loses on its trip out the electrical signal will gain on its trip back, thus, cancelling out any expected interval.

---

mechanical clock) whereby eclipses of the moons of Jupiter could be used to detect a motion of the solar system through the ether in much the same way as the experiment described here.

2-7 THE MICHELSON-MORLEY EXPERIMENT

The fact that light reaches all walls of a room shows that the earth is not moving through the ether with a speed greater than c. Careful measurements of the intensity of light going in opposite directions from a uniform light source showed that, in fact, it cannot be moving through it at any large fraction of the speed c. The possibility always remained that the motion of the earth through the ether, was just too small to be detected until Michelson developed his interferometer, and performed the Michelson-Morley experiment, which was sensitive enough to detect the orbital motion of the earth about the sun.

The Michelson-Morley apparatus is shown, in principle, in Figure 2-9. A source S sends out a flash of light that travels on two paths at right angles to each other, one to the mirror A and back to S, the other to the mirror B and back. For simplicity we assume that the distance L to each mirror is the same. If the apparatus were at rest with respect to the ether, the two flashes would each take a time $2L/c$ for the round trip. Suppose, however, that the whole apparatus is in motion through the ether toward the right as in Figure 2-10. The light flash leaves when the source is at S_1. By the

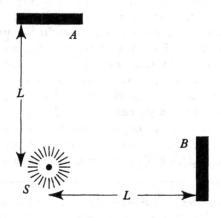

Figure 2-9

Light Waves and the Second Postulate

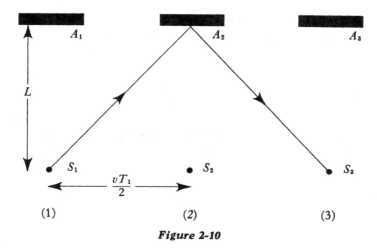

Figure 2-10

time it has reached the mirror A at A_2, the apparatus has advanced to the position 2, and the reflected flash returns to the source at S_3. Suppose the whole process takes a time T_1. Then, if the apparatus is moving through the ether with a speed v, the distance from S_1 to S_2 is $vT_1/2$. The path the light follows is

$$S_1A_2S_3 = 2\sqrt{L^2 + \left(\frac{vT_1}{2}\right)^2} \tag{2-6}$$

Since the light is presumed to travel with a speed c with respect to the ether, we have

$$T_1 = \frac{S_1A_2S_3}{c} = \frac{2}{c}\sqrt{L^2 + \left(\frac{vT_1}{2}\right)^2} \tag{2-7}$$

Solving for T_1 we have

$$T_1 = \frac{\frac{2L}{c}}{\sqrt{1 - \frac{v^2}{c^2}}} \tag{2-8}$$

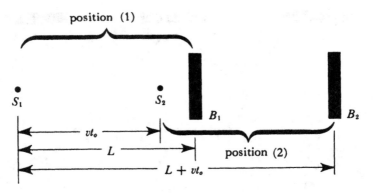

Figure 2-11

Figure 2-11 shows the other arm as it moves through the ether. Suppose the trip out to the mirror B takes a time t_o. Then by the time the light has reached the mirror it has advanced a distance vt_o. The total distance the light went in the time t_o is therefore $L + vt_o$. Since it traveled at a speed c, we have

$$ct_o = L + vt_o$$

and

$$t_o = \frac{\frac{L}{c}}{1 - \frac{v}{c}} \tag{2-9}$$

Similarly the trip back takes a time t_b where

$$t_b = \frac{\frac{L}{c}}{1 + \frac{v}{c}} \tag{2-10}$$

Light Waves and the Second Postulate

The total time $T_2 = t_o + t_b$ is

$$T_2 = \frac{\frac{2L}{c}}{1 - \frac{v^2}{c^2}} \tag{2-11}$$

The times T_1 and T_2 are not the same, and if the difference in time of arrival of the return beams could be measured, the speed of the apparatus through the ether could be determined.

As it has been described, such a measurement is hopeless unless the speed of the earth through the ether is very great. As actually performed, the experiment made direct use of the fact that light is a regular wave train of periodic waves. The actual arrangement of the apparatus is shown in Figure 2-12. The source of light S is

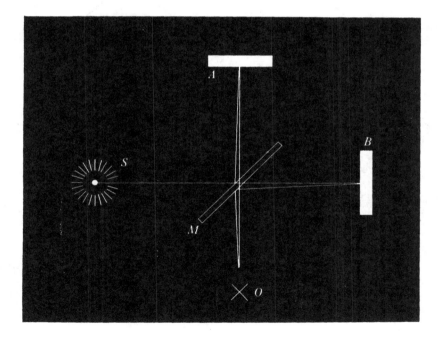

Figure 2-12

actually on continuously. It shines on a half-silvered mirror M. Half the light is reflected to the mirror A and half passes through to the mirror B. Upon its return, half the light from A passes through M to the observer at O, and half the light from B is reflected to O. The observer therefore sees a superposition of two light beams, one of which has gone along $SMAMO$ and the other $SMBMO$.

Remember that light is an electromagnetic wave, and a regular, periodic light wave produces an oscillating electric field at the

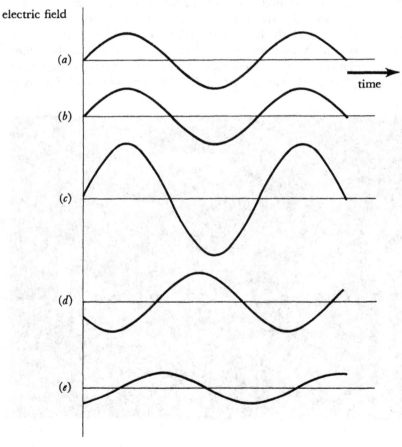

Figure 2-13

position of the observer's eye. A graph of such an electric field as a function of time is shown in Figure 2-13a. Suppose this is the wave received along the path *SMAMO*. If the wave received along *SMBMO* takes exactly the same time to arrive, its oscillations, Figure 2-13b, are the same as those of Figure 2-13a; the sum is shown at (c). So, if the two paths are of identical lengths, the observer sees a bright image. If, on the other hand, the light from *SMBMO* takes longer to reach the observer, the wave is delayed as at (d). The sum is shown at (e). Clearly the observer sees a less intense image. If the *SMBMO* beam is delayed by one-half period with respect to *SMAMO*, the sum is zero and the observer sees darkness. For instance, if the mirror B is moved away from M, the intensity of the light seen at O gets alternately bright and dark as the time of arrival is successively delayed by one-half, one, one-and-a-half, two, etc., periods. In actual practice, the mirror at B is tilted slightly, so that the image gives the appearance of light and dark stripes, or fringes. A time delay in one branch will shift the fringes sideways. The actual apparatus used had a path $L = 11$ meters which was actually produced by multiple reflections between a series of mirrors in a device about $1\frac{1}{2}$ meters across. The whole apparatus was floated on a pool of mercury so that it could be rotated slowly. Since the light was supposed to take longer to return along the path parallel to the motion, the fringe system should slowly shift back and forth as first one and then the other arm became parallel to the earth's motion through the ether. No such shift was observed!

The experiment was designed to be sensitive enough to detect the orbital motion of the earth about the sun (see Exercise 3); yet no shift of the fringes was observed! There are several possible explanations for the effect—all of them rather radical in nature.

1. The earth "drags" the ether along with it and, hence, the apparatus fixed to the earth is not moving through the ether. This is the explanation proposed by Michelson himself. However, it is not tenable, since it would also lead to no observation of the aberration of starlight, and as we have seen this is observed.

2. The arm of the apparatus which is moving parallel to the motion through the ether contracts by a factor $\sqrt{1 - v^2/c^2}$ and exactly cancels the expected shift. By reference to Equation 2-11 it can be seen that such a contraction would exactly cancel the time

difference expected in the Michelson-Morley experiment. If L were replaced by $L\sqrt{1 - v^2/c^2}$, Equation 2-11 would become identical to Equation 2-8, and there would be no time difference expected. This explanation, the so-called Lorentz-Fitzgerald contraction, is not quite so arbitrary as it may seem at first sight. If all the forces that hold the atoms of the apparatus together were electrical in nature, such a change in force, and hence length of arm, would be expected. As we shall see, such a contraction is also predicted in the theory of relativity, and many of the ideas of relativity were already implicit in electromagnetic theory—though not, of course, recognized as such. The performance of the Michelson-Morley experiment with an apparatus whose arms were of quite unequal length subsequently disproved the limited explanation provided by the Lorentz-Fitzgerald contraction.

3. The theory of electromagnetism should be modified. A number of attempts were made to alter the generally accepted theory of electromagnetism. All of these are too complicated to be detailed here. Also they were all very unsatisfying and, what is worse, internally inconsistent. [4]

4. Light is not a wave in the ether, but rather composed of particles subject to the laws of Newtonian mechanics. The reader should prove to himself (see Exercise 1) that if light were composed of particles which have a particular speed with respect to the source and not with respect to a medium, there would be no time delay observed in the Michelson-Morley apparatus. The difficulty with this explanation is that light *is* a wave motion. The fringes observed with the Michelson-Morley apparatus are themselves beautiful proof of its wave nature. But the words "subject to the laws of Newtonian mechanics" suggest the radical way out chosen by Einstein. What "laws of mechanics" is it necessary that light obey in order to explain the results of the Michelson-Morley experiment? Only the principle of relativity. If the apparatus is at rest with respect to the ether, the light takes equal times to traverse the two paths of the apparatus. *If* we could apply the principle of relativity,

[4] The extent of the evidence against the most complete of these theories, that proposed by Ritz, is discussed by J. G. Fox, *Am. J. Phys.* **33**, 1 (1965). This article includes an interesting critique of the double star argument presented in Section 2-9.

Light Waves and the Second Postulate

we would find the same result in any inertial frame of reference, i.e., with the apparatus moving uniformly with respect to the ether. This is, of course, just the result of the Michelson-Morley experiment. Einstein understood the Michelson-Morley experiment in just these basic terms and proposed that:

5. Light, and in fact all electromagnetic phenomena, obey the principle of relativity. We now believe that this is the proper "explanation" of the Michelson-Morley experiment. At the turn of the century, mechanics and electromagnetic theory encompassed almost all known physics, and it was not a very great step to propose the first postulate of relativity as applicable to all physical phenomena.

2-8 THE SECOND POSTULATE

If light were a stream of mechanical particles that obeyed mechanical laws, there would be no difficulty in understanding the results of the Michelson-Morley experiment (as the reader is to show in Exercise 1).

Suppose, for example, that a rocket ship were traveling at a speed $(\frac{1}{2})c$ with respect to some observer and a beam of light were shone ahead of it from the nose. If the speed of light meant the speed of the light "particles" with respect to their source, then these light "particles" would be traveling at a speed $c + (\frac{1}{2})c = (\frac{3}{2})c$ with respect to the observer. But that kind of behavior is very unwavelike, for waves travel at a certain speed with respect to the medium in which they propagate, and not with a certain speed with respect to the source. And if one thing is clear from the many beautiful experiments of optics, and indeed from the whole theoretical structure of electromagnetism, it is that light is a wave phenomenon. This conclusion seems so certain that Einstein made this property of a wave the basis of his second postulate:

The speed of light is independent of the speed of its source.

Now this conclusion would not be at all remarkable when taken by itself. It only states that light behaves like a wave; it might, from this postulate alone, have a speed c with respect to some medium, the ether. It is when this postulate is combined with the first postulate that remarkable conclusions begin to emerge.

Return for a moment to the simple (in principle) detectors and flasher of Figure 2-7. Some experimenter could flash the light and measure the difference in time of arrival of the two flashes. Suppose he found zero for the time difference; this would merely indicate he was at rest with respect to the ether. A second observer, moving uniformly past the first, carrying a similar set of detectors and flasher would necessarily be moving with respect to the ether; he should find a time difference according to Equation 2-5. But this would mean that two uniformly moving observers would have performed identical experiments and obtained different answers, and *this is contrary to the postulate of relativity*. No one has yet performed the simple experiment described, but many physicists have performed the Michelson-Morley experiment, and its conclusions amount to the same thing. In fact, many ingenious electromagnetic experiments have been performed; none contradicts the hypothesis that two uniformly moving observers get the same results when they perform the same experiment. Also many ingenious electromagnetic and optical experiments have been performed all of which attest the fact that light is an electromagnetic wave—and the speed of a wave is independent of the speed of its source.

2-9 EXPERIMENTAL PROOF OF THE SECOND POSTULATE

We must emphasize that at the time Einstein proposed it, there was no direct experimental evidence whatever for the speed of light being independent of the speed of its source. He postulated it out of logical necessity.

An experimental verification of the second postulate would seem to depend on the measurement of the speed of light from a rapidly moving source. If the speed is independent of the speed of the source, then the second postulate is verified. In principle it seems easy to verify the statement, and yet the author knows of no completely direct experimental evidence. The most nearly direct evidence comes

Light Waves and the Second Postulate

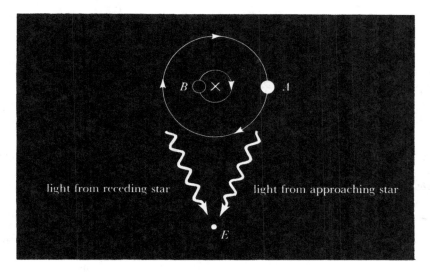

Figure 2-14

from three sources: (1) observations of double stars in a way related to that originally proposed by De Sitter [5]; (2) measurement of the speed of the radiation from the annihilation of rapidly moving positrons [6]; (3) measurement of the speed of the photons from the decay of rapidly moving π^0 mesons. [7]

Figure 2-14 shows a double star system with two stars, A and B, rotating about their center of mass. An observer on the earth at E is able to measure the speed at which the star A is approaching or receding by measuring the Doppler change in wavelength or frequency of certain spectral lines in the light from the star.[8] Suppose the double star were nearby. The measured speed of approach would

---

[5] De Sitter, *Proc. Amsterdam Acad.*, **16,** 395 (1913).

[6] Sadeh, *Phys. Rev. Letters*, **10,** 271 (1963).

[7] Alväger, Farley, Kjellman, and Wallin, *Phys. Letters*, **12,** 260 (1964).

[8] It might be argued that if we are uncertain about the nature of light and its speed, how do we know how to measure a speed of the source by the Doppler shift in frequency? The answer is that the Doppler shift is observed directly for terrestrial sources, and the effect can be calibrated as a speedometer independent of any theory.

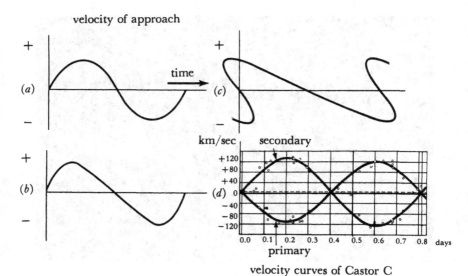

Figure 2-15

look like the graph in Figure 2-15a. Now, suppose the star were farther away and, furthermore, the speed of light coming from the star were increased when the star was approaching and decreased when the star was receding. Then the light from the star would reach earth relatively sooner when it came from the approaching part of the orbit and later from the receding part of the orbit. The *measured* approaching and receding curve would look like Figure 2-15b. In fact, the curve could even become distorted as in Figure 2-15c where the light from one approaching part of the orbit reaches the earth *before* light from the *previous* receding part! Figure 2-15d shows an actual measurement on the binary star Castor C.[9] The graph shows the curves for both members of the binary; as one star approaches the earth, the other recedes from it. There is no evidence for any distortion in the curves, and none has been seen which is not attributable to eccentricities in the orbits. The absence of such distortion indicates that the speed of light does not depend on whether the

[9] A. H. Joy and R. F. Sanford, *Astrophys. J.*, **64**, 250 (1926).

Light Waves and the Second Postulate

source is approaching or receding, i.e., its behavior is that expected of a wave motion.

A second rather direct proof that the speed of light is independent of that of the source, comes from the annihilation of rapidly moving positrons. A positron is essentially a positively charged electron that is given off from many radioactive substances, and is produced in many large particle accelerators. When a positron collides with an electron, the two particles are sometimes annihilated, i.e., the particles disappear and two photons are given off. The photons are essentially flashes of electromagnetic radiation, i.e., light. Now it is possible to produce positrons that have a speed that is a very large fraction of the speed of light. When these positrons hit electrons at rest in a material, occasionally one of the photons goes forward and the other backwards. If the speed of the photons depended on the speed of their source, the one going forward would travel faster than the one going to the rear. If two detectors were placed at equal distances from the place where the annihilation occurred, the forward detector should detect the photon first. When the experiment is performed, no time difference is observed, indicating that the photons travel with equal speeds.

Simple interpretation of this latter experiment is made somewhat troublesome by the fact that the "source" is composed of one moving particle and one particle nearly at rest. The double star experiment has also been called into question on the grounds that we do not see radiation direct from the star, but rather radiation that has been absorbed and reradiated by a gas cloud around the double star system.

The third test is the most nearly direct. Neutral pions (π^0) are produced in large numbers by very high-energy accelerators. These mesons live only about 10^{-16} sec and then decay into two photons. In the experiment described, it was possible to observe only those photons emitted in the forward direction by π^0-mesons traveling faster than 99.98 percent of the speed of light. Very short bursts of these photons were produced and directly timed over a flight path of 31 meters. The measured speed of these photons was $2.9977 \pm .0004 \times 10^8$ meters/sec, in excellent agreement with the speed of light (2.9979×10^8 meters/sec) as measured from a stationary source. This experiment seems open to almost no question. However, it should be remembered that experimental proof for the postulates of relativity rests not so much on the direct evidence presented here, but on the tremendous weight of indirect evidence for the logical consequences of the two postulates that will be considered in subsequent chapters.

2-10 A COROLLARY TO THE SECOND POSTULATE

There is an alternative way of phrasing the second postulate which is sometimes easier to apply directly to problem solving than the one given in the Introduction. If the speed of light is independent of the speed of the source, then in a particular reference system the speed of light has the same value from moving as from stationary sources. In a second frame of reference, moving uniformly with respect to the first, the same statement is true. Measuring the speed of light from a stationary source is an experiment which must give the same result in both reference frames. It therefore follows that:

The speed of light is the same in all inertial frames of reference.

This speed of light we shall denote by c. No matter which frame of reference is chosen for the solution of a problem, the speed of light must be taken as c. This is somewhat equivalent to saying that the medium in which the light wave travels is at rest with respect to any inertial reference frame. If such a concept is helpful, very good, but it is apparent that the ether no longer plays any essential role in our theory of light. There is no longer any reason to suppose it exists; therefore, we shall no longer use the concept in this book.

Exercises

1. Assume that the source S in the Michelson-Morley apparatus emits particles which travel out from the source with a speed c. In the frame of reference where the apparatus is at rest, the round trip from source to mirror to source takes a time $2L/c$. Repeat this calculation in the frame of reference where the apparatus is moving to the right with a speed v, i.e., the cases shown in Figure 2-11a and b. Calculate explicitly the velocities of the particles in this frame of reference and then get the times of flight.

Light Waves and the Second Postulate

2. In the text, the expected result of the Michelson-Morley experiment was calculated in the frame of reference where the *medium*, the ether, was at rest and the apparatus moving. Assuming that light is a wave, calculate the time for a round trip over both arms, but do the calculation in a frame of reference where the apparatus is at rest and the medium is moving to the left with a speed v.

3. The path L of the Michelson-Morley apparatus was actually 11 meters long. The wavelength of the light used was 5.9×10^{-7} meters. What time delay between the two arms would delay the light from one arm by $\frac{1}{2}$ period with respect to the other arm? A time delay (fringe shift) of about 1 percent of this could have been detected. What velocity of the apparatus through the ether would give rise to this latter time delay? (You will probably have to use the results of Appendix A.) What is the velocity of the earth in its orbit?

4. A certain spectral line in the spectrum of a double star ordinarily has a frequency of 0.5×10^{15} cycles/sec. If the star is approaching the earth at 200 km/sec, approximately by what fraction $\Delta f/f$ will the frequency be changed?

5. The star Castor C, whose velocity curve is shown in Figure 2-15d, is 45 light years away. At zero time on the graph the light shows no Doppler shift, which means that the star was neither approaching nor receding from the earth. This light traveled at the speed c and therefore took 45 years to arrive at the earth. Light emitted 0.2 days later was emitted when the star was traveling at a speed 130 km/sec toward the earth. Suppose that this light traveled at a speed $c + 130$ km/sec. How long would it take to arrive at the earth? It would arrive, not just 0.2 days after the Doppler unshifted light, but at some earlier time. When? In this way construct the curve, like Figure 2-15b or 2-15c, which would represent light from this star *if light had a speed c with respect to its source*. For this problem use the velocity curve for the star labeled "secondary." Why is there a small shift of the horizontal axis toward positive velocities?

Answers to Completion Sentences on Pages 25 and 26

a. $\dfrac{r_1}{c}$

b. $\dfrac{1}{f_s}$

c. $\dfrac{1}{f_s} + \dfrac{r_2}{c}$

d. $\dfrac{1}{f_s} + \dfrac{r_2}{c} - \dfrac{r_1}{c}$

e. $r_2 = \sqrt{r_1{}^2 + \overline{O_1O_2}{}^2 - 2r_1\overline{O_1O_2}\cos\theta}$

f. $r_2 = r_1 - \overline{O_1O_2}\cos\theta$

g. $\overline{O_1O_2} = \dfrac{v}{f_o} = v\Delta t$ (not f_s)   Then we have

$$\Delta t = \dfrac{1}{f_s} + \dfrac{r_1 - \overline{O_1O_2}\cos\theta}{c} - \dfrac{r_1}{c} = \dfrac{1}{f_s} - \dfrac{v\Delta t}{c}\cos\theta$$

Solving, $\Delta t = \dfrac{1}{f_s\left(1 + \dfrac{v}{c}\cos\theta\right)}$ etc.

3

Time Dilation: Proper and Improper Time

3-1 GEDANKEN EXPERIMENTS

THE MOST IMPORTANT EXPERIMENT that demanded a revision of the classical ideas about the ether and the propagation of light, and which led to Einstein's two postulates of relativity was the Michelson-Morley experiment. It is reasonable, then, to start our discussion of the consequences of these postulates by considering a somewhat idealized version of an experiment with the Michelson-Morley apparatus. The apparatus will consist of a frame upon which we mount a light flasher, a mirror, and a clock. We are going to measure the time it takes for the light to leave the flasher, travel to the mirror, and return to the clock.

It is worthwhile to say a word or two in general about our proposed "experiment." We are not going to quote any real results from any real experiment done with an apparatus that even remotely resembles the one described, because, if the clock mentioned were an ordinary one with a second hand, and we were prepared to measure time intervals of only a few seconds, the frame would have to be hundreds of thousands of miles long—clearly a technical impossibility. On the other hand we might use rather more sophisticated clocks—a sweep on a fast oscilloscope, for instance—so that time intervals of a few nanoseconds (10^{-9} sec) could be easily measured and the apparatus would only have to be a few meters long. This would help very little since we shall see shortly that our experiments require us to move the apparatus at speeds comparable to the speed of light. Even in this day and age where interplanetary travel seems just around the corner, such speeds are unattainable with apparatus larger than atomic size. The apparatus we have just described will be used by us then in thought only; it is, therefore, another "gedanken" experiment. It is only a crutch to help us in our thinking about the logical consequences of certain postulates. However, we can easily *imagine* frameworks the size of the solar system moving with speeds approaching that of light; it is "merely" a bit of engineering that prevents us from actually carrying out our gedanken experiments. Fortunately, there are experiments that can actually be performed, which tell us that our postulates are correct; but, unfortunately, most of them are indirect and not suitable for a first look at the logical consequences of our postulates.

3-2 MEASUREMENT OF PROPER AND IMPROPER TIME INTERVALS

In this chapter we will consider the arm of the Michelson-Morley apparatus that moves perpendicular to its length. Figure 3-1 shows the idealized apparatus. It consists of a rigid frame on which is mounted a light source S and a mirror A. These are placed a distance L apart. A clock rests beside the light source. The source emits a flash of light which illuminates the clock and also travels out to the mirror and back, where it again illuminates the clock. We may think of an observer who observes the travel of the clock hands and can therefore measure with this clock the time the light flash takes to go the distance $2L$ to the mirror and back. Sometimes it may be helpful to have this "observer" be a camera mounted over the clock, which records a permanent picture of the clock face that may then be studied at leisure.

Referring to Figure 3-1 we see that the light flash travels a distance $2L$. The second postulate demands that it travel at a speed c and,

Time Dilation: Proper and Improper Time

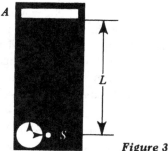

Figure 3-1

therefore, that the time taken for the round trip be $2L/c$. The photograph of the clock will show that the hands have advanced by a time $2L/c$ between flashes. Although it may seem overly fussy in this simple case, let us repeat this in more precise terminology. We have measured the *time interval* between *two events*: (1) the sending out of the signal and (2) the receipt of the return signal. In particular we have measured this time interval on a clock that was present at both events; the signal started from the clock and arrived back at the same clock. Such a time interval is called a *proper time interval*.

We now consider the *same experiment* from the point of view of a different observer, one who is moving to the left in Figure 3-1 with a speed v. In this observer's frame of reference the apparatus is moving to the right with a speed v (cf. Section 1-6) as shown in Figure 3-2.

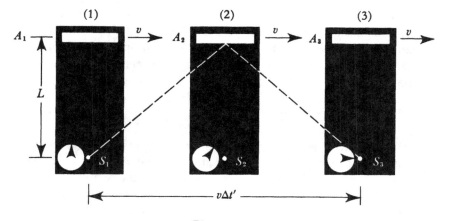

Figure 3-2

Suppose the flash of light leaves the source and illuminates the clock carried by the apparatus at S_1. The light travels to the mirror and back to the source, but during this time the source and mirror have moved from S_1A_1 to S_2A_2 and finally to S_3A_3. The light has traveled from S_1 to A_2 to S_3 as shown in Figure 3-2. Suppose this trip took a time $\Delta t'$. (We use the symbol Δt to emphasize that we are considering a time *interval*.) Then the distance S_1S_3 must be $v\Delta t'$. The distance $S_1A_2 = \sqrt{L^2 + (v\Delta t'/2)^2}$ and the whole distance $S_1A_2S_3 = 2\sqrt{L^2 + (v\Delta t'/2)^2}$. This frame of reference is also an inertial frame of reference because it is moving uniformly with respect to our first one. Therefore light travels at a speed c and we have

$$c\Delta t' = 2\sqrt{L^2 + \left(\frac{v\Delta t'}{2}\right)^2}$$

Solving for $\Delta t'$ gives

$$\Delta t' = \frac{\dfrac{2L}{c}}{\sqrt{1 - \dfrac{v^2}{c^2}}}$$

In our rather formal language, we have measured the time interval between the sending and receipt of the light flash, the *same two events* as in the first case, but this time in a frame of reference where the two events occur at different points, i.e., could not be measured by the same clock. This time interval is called an *improper time interval* between these two events. Since $2L/c$ is the proper time interval, the proper and improper time intervals between two events are related by

$$\Delta t(\text{proper}) = \Delta t(\text{improper})\sqrt{1 - \frac{v^2}{c^2}} \tag{3-1}$$

Therefore, we have the first really outstanding result of the special theory of relativity, namely, that the time interval between two events as seen by different observers depends on the relative velocity of those observers. This effect is called *time dilation* or *time dilatation*.

This result is so contrary to our usual use of clocks and understand-

Time Dilation: Proper and Improper Time

ing of the nature of time that we should consider in detail how we might make the measurement in this second frame of reference where the apparatus is moving. Consider Figure 3-3. We have provided our second observer with a set of clocks, each identical to the one carried on the apparatus. This entire set of clocks is synchronized, i.e., set to run together. This may be done, in principle, by getting all the clocks together, setting them, and then carrying them out to their appropriate locations. If one were actually to do an experiment even remotely resembling this one, the clocks would be placed at measured positions and a light or electric signal sent from one clock to all the rest. The fact that light travels at a speed c allows an accurate correction for the time delay in arrival of the signal at the various clocks.

As the apparatus moves past the line of clocks, the source sends out its flash of light at S_1 illuminating both the clock carried on the apparatus and the stationary clock past which it happens to be traveling, clock 3 in Figure 3-3. The apparatus moves down the line of clocks and the return flash happens to illuminate clock 14. The speed of the apparatus, v, can be conveniently determined by measuring the distance between clock 14 and clock 3, noting the time interval $\Delta t'$ between the reading of clock 14 and clock 3, and dividing.

The reader must now realize that there is nothing imaginary in

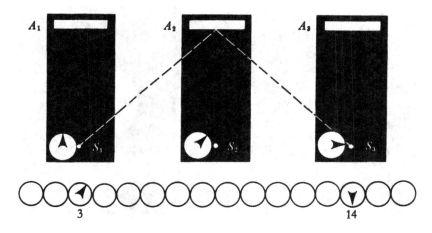

Figure 3-3

this time difference as measured by these various observers. If photographs of the clocks were taken by the outgoing and returning light flash, they could be brought back and compared and would look like the clocks in Figure 3-4. The clock carried on the apparatus reads a time lapse of 15 min between the two pictures of the minute hand on the same clock. The time difference between the two clocks past which the apparatus was moving is 25 min. (The reader should now calculate how fast the apparatus was moving past the stationary clocks. See page 62 for answer.)

3-3 DO MOVING CLOCKS RUN SLOW?

Figure 3-4 shows that the clock on the apparatus, the one most people would think of as the "moving clock," measures a smaller interval between the two events. It is sometimes stated, in simple terms, that "moving clocks run slow." There is nothing wrong in this statement, but it gives rise to an apparent paradox due to muddle-headed thinking: If a clock A is moving past another clock B, then A runs slower than B. But from A's point of view, B is moving past A and, therefore, B runs slower than A. This is a logical contradiction and often leads to even worse ideas about relativity, such as: Since clocks A and B cannot both run slow with respect to each other, it

Figure 3-4

Time Dilation: Proper and Improper Time

must be that A only *seems* to run slow to someone standing by B, and that B only *seems* to run slow to an observer at A. This is often erroneously attributed to the fact that light takes some time to get from the clock A to the observer at B and that is why A seems to run slow. Nothing could malign the theory of relativity more thoroughly. The difference in readings on the clocks is really there. As implied by Figure 3-4, it could be photographed and the photographs studied by someone standing on the apparatus as well as someone on the ground; they would both agree on the results. The key to resolving the apparent paradox is to consider carefully what it is that clocks measure. Clocks do not tick on and on, measuring some mystical quantity called *Time*. They are mechanical devices that can be started and stopped (or photographed) in order to measure intervals between events. In the instance considered here, there were two events, the flash and the return of the flash. In only one frame of reference may the interval between these events be measured directly by a single clock—the frame of reference of the apparatus. In all other frames, the interval appears longer. The minimum time interval between these events occurs in a unique frame of reference. The clock measuring the proper time interval between two events "goes the slowest." There is no paradox here.

But most readers will doubtless still ask about placing a similar flasher-mirror apparatus on the ground beside clock 3, for example, as shown in Figure 3-5. A flash sent out from this apparatus will return to clock 3. Clock 3 will, therefore, measure a proper time interval between the sending and receipt of this new flash, and will measure a smaller interval between these new events than clocks fixed on the first apparatus. It appears the paradox has returned. Here is clock 3 going faster and slower than clocks on the first apparatus. But, there is no logical contradiction because this new pair of events is *different* from the first pair. The return of the first light flash was to clock 14; the second return considered was to clock 3—clearly a different event. Whether or not clocks go slow or fast compared to clocks moving with respect to them must be related to the particular events concerned, and not to some nebulous concept called absolute time.

In summary, then, it is all right to say "moving clocks run slow" if you know what you mean, but a much more meaningful and

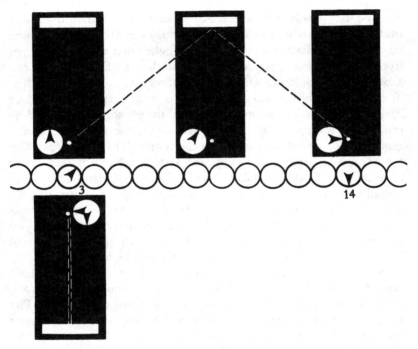

Figure 3-5. *This identical apparatus is stationary with respect to the line of clocks. What would this look like if drawn in the frame of reference where the original apparatus is stationary?*

precise statement is: "A clock measuring a proper time interval between two events (i.e., present at both events) measures a smaller interval than clocks measuring an improper interval between these same events."

3-4 IS THE CLOCK MECHANISM AFFECTED BY MOTION?

Some readers may look at Figure 3-4, see that the "moving clock runs slow" and, therefore, feel that somehow the rapid movement has affected the works of the clock, thereby making it "really run slow." Such a viewpoint is inconsistent with the first postulate of

Time Dilation: Proper and Improper Time

relativity. The two uniformly moving systems are both inertial systems. All physical laws in them, including mechanics, are the same. Therefore, the mechanisms of identical clocks run in the same way in both systems. The motion does not affect the clocks in the slightest. The fact that the time interval between two events is different in the moving systems is a kinematical one, much more closely connected with what we mean by a time interval than with the works of a clock.

3-5 REAL EXPERIMENTS IN TIME DILATION

The result of the previous section is so startling that no one can be expected to be satisfied with a gedanken experiment. The first observation of an effect closely related to that of time dilation was the experiment of Ives and Stilwell [1] who measured the change in frequency of the spectral lines emitted by rapidly moving atoms and found perfect agreement with the theory of relativity. The effect, though accurately measured, was quite small since the velocities attained by the moving atoms were only $\frac{1}{2}$ percent that of light. Really dramatic effects are afforded by the observation of time dilation in μ-mesons, or muons, in cosmic rays. These effects were first observed by Rossi and Hall.[2] We will describe here the results of a more recent simplified version of their experiment.[3]

Muons are charged particles that have a mass about 200 times that of an electron. They decay spontaneously into an electron, a neutrino, and an anti-neutrino,[4] with a half-life of 1.53×10^{-6} sec. Although it is uncertain when any particular muon will decay, accurate statistical predictions can be made. For instance, if 1024 muons are present at any one time, then on the average 512 will be present after 1.53×10^{-6} sec have passed, 256 after 3.06×10^{-6} sec, 128 after 4.59×10^{-6} sec, etc. These results are subject to random sta-

---

[1] H. E. Ives and G. R. Stilwell, "An Experimental Study of the Rate of a Moving Atomic Clock," *J. Opt. Soc. Am.*, **XXVIII**, 215 (1938). See also Chapter 5, page 90, of this book.

[2] B. Rossi and D. B. Hall, *Phys. Rev.*, **59**, 223 (1941).

[3] D. H. Frisch and J. H. Smith, *Am. J. Phys.*, **31**, 342 (1963).

[4] A neutrino is a massless, electrically neutral particle involved in radioactive decays. In muon decay, recent evidence indicates that there are actually two kinds of neutrinos involved. Such details do not affect the argument given here.

tistical fluctuations. The times are measured, of course, in the frame of reference where the muons are at rest.

A group of muons can thus be used as a convenient clock for measuring short time intervals. If 568 muons are present at one event, but by the time a second event has occurred only 142 are left, two half-lives, or 3.06×10^{-6} seconds, have passed.

Muons are the most abundant particles present in cosmic rays at altitudes of a few thousand feet and are mostly traveling vertically downward at speeds close to that of light. The experiment consists of measuring the time of flight of muons over a path of a few thousand feet by means of the muon clocks themselves and by clocks fixed to the earth. A counter A in Figure 3-6 was installed on top of Mt. Washington, New Hampshire, at an altitude of 6265 ft. It was adjusted to count muons with speeds between $.9950c$ and $.9954c$, i.e., 99.52 percent the speed of light, and it was found that 563 ± 10 muons in this speed range arrived in the apparatus on Mt. Washington each hour. When the same apparatus was removed to a location only 10 ft above sea level, it was found that 408 ± 9 muons/hr arrived in it. In the actual experiment, therefore, the same muons

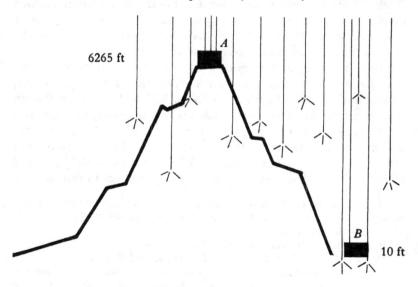

Figure 3-6

Time Dilation: Proper and Improper Time

were not actually timed between 6265 ft and 10 ft, but since cosmic ray intensities are known not to fluctuate from time to time and place to place—at least not over the distance our apparatus was moved—the results are the same as if one group of muons had been timed.

Let us calculate the flight time, which would be measured with clocks fixed to the earth. The muons were actually slowed down appreciably by the air in their flight so that their average speed was about $.992c$. The flight time was, therefore,

$$T_{im} = \frac{(6265 - 10) \text{ ft}}{.992 \times 3 \times 10^8 \frac{\text{meters}}{\text{sec}} \times 3.28 \frac{\text{ft}}{\text{meters}}} = 6.4 \times 10^{-6} \text{ sec}$$

The subscript im is used to designate T, since this is an improper time interval for the events in question. The events are: (1) muon (or muons) passes Mt. Washington; (2) muon arrives at sea level. It is the muon clocks that are present at both events and measure the proper time interval.

Since many more than half the muons live to reach sea level, the time measured by the muon clocks is certainly less than 1.5×10^{-6} sec. A more accurate estimate can be obtained by remembering that in a random decay process starting with N_0 particles, after a time t there are N left, where

$$N = N_0 e^{-(t/t_0)} \qquad (3\text{-}2)$$

The time t_0, called the mean-life of the particles, is related to the half-life. When half the particles are left $N = N_0/2$ and t must be the half-life, $t_{\frac{1}{2}}$. Therefore,

$$\frac{1}{2} = e^{-(t_{\frac{1}{2}}/t_0)}; \qquad -(t_{\frac{1}{2}}/t_0) = \log_e \frac{1}{2}; \qquad t_0 = t_{\frac{1}{2}}/\log_e 2$$
$$= 1.53 \times 10^{-6}/.693$$
$$= 2.21 \times 10^{-6} \text{ sec}$$

The flight time from Mt. Washington to sea level is such that

$$408 = 563 e^{-(t/2.21 \times 10^{-6})}; \qquad t = .715 \times 10^{-6} \text{ sec}$$

as measured by the muon clocks. Thus the proper time interval for the trip from the mountain to sea level is

$$T_p = .715 \times 10^{-6} \text{ sec}$$

There is no doubt that there is a startling difference in these values!
In the experiment, the ratio of proper to improper times was

$$\frac{T_p}{T_{im}} = \frac{.715 \times 10^{-6}}{6.4 \times 10^{-6}} = .11$$

The ratio predicted by Equation 3-1 is

$$\sqrt{1 - \frac{v^2}{c^2}} = \sqrt{1 - .992^2} = .13$$

The agreement is well within the experimental errors. The description here has been necessarily brief; the interested reader is urged to read the original article for details, including the determination of the speed range.

One final word: Notice that the effect measured in this experiment was not tiny. The muons were observed to decay at only one-ninth of the rate at which they decay at rest. Their "clocks" were "running slow" by a factor of nine! Every day, high-energy nuclear physicists working at high-energy accelerators work with beams of other particles, pions and kaons, which decay spontaneously over 100 times as fast as muons. Were it not for time dilation, they would decay and disappear before they had traveled several feet, even traveling at the speed of light. Because their decay is slowed down, they can be observed hundreds of feet from the point in the accelerator where they are produced. Consequently, they can be usefully employed in further experiments. Time dilation, therefore, becomes a matter of everyday concern to these physicists.

Exercises

1. A beam of K^+ mesons passes through three counters spaced 9 meters apart as shown. K^+ mesons decay radioactively.

Time Dilation: Proper and Improper Time

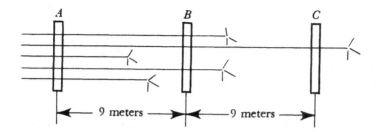

a. If 1000 K^+ mesons go through counter A, and 250 live to arrive at counter B, approximately how many will get to counter C?

b. If the speed of these mesons is 86.6 percent of the speed of light, what is the half-life of a K^+ meson measured in a frame of reference where it is at rest?

2. A π^0 meson is an uncharged π meson which decays into two photons (electromagnetic radiation) with a half-life of about 2×10^{-16} sec. If a π^0 meson were produced at the nucleus of an atom, how fast would it have to be going to leave the atom in which it was produced in one half-life? how fast to go 10^{-5} meters? how fast to go a foot? (Assume an atom is 10^{-10} meters in radius.)

3. A satellite orbits the earth at a height of a few hundred miles in about 100 min. By how many seconds per day will a clock in such a satellite run slow compared to an earth clock? (Compute just the special relativistic time dilation. There are other, nearly comparable, gravitational effects.)

4. Alpha Centauri is a star about 4 light-years away. For a rocket ship to make the trip in one day—as reckoned by its occupants—it would be necessary to go how fast? To the rocket ship's occupants, α-Centauri would appear to be approaching at this same speed. Therefore, how far away would it appear, to them, to be at the start of the trip?

5. A rocket leaves Earth (event A), flies to Mars (event B), and sends back a light signal which arrives on earth (event C).

a. Whose clocks are proper clocks for events $A \to B$?
b. Whose clocks are proper clocks for events $A \to C$?
c. Whose clocks are proper clocks for events $B \to C$?
d. What is the proper time interval between events B and C?
e. With what speed did the return light-flash travel with respect to the rocket?
f. With what speed did the return light-flash travel with respect to the earth?

6. A rocket ship leaves the earth at a speed $(\tfrac{3}{5})c$. When a clock on the rocket says one hour has elapsed, the rocket sends a light signal back to Earth.

a. According to *earth clocks* when was the signal sent?

b. According to *earth clocks* how long after the rocket left did the signal arrive back on earth?

c. According to *rocket clocks* how long after the rocket left did the signal arrive back on earth?

d. Whose clocks run slow?

e. It is instructive to work parts *b* and *c* in both earth and rocket reference frames and show that the answers agree.

7. A rocket ship leaves the earth at a speed of $(\frac{12}{13})c$. The rocket reports back by radio every hour according to *its* clocks. What will be the time interval between reports received back on Earth? If the rocket returns at the same speed directly toward earth and continues to report as before, what will be the new time interval between receipt of reports?

Answer to Question on Page 54

$$v = \frac{4}{5}c$$

4

Length Measurements

4-1 LENGTH CONTRACTION

FIGURE 4-1 SHOWS A ROCKET SHIP traveling between two markers a distance L apart. In the frame of reference of the markers the rocket ship is traveling at a speed v. It therefore covers the distance L in a time L/v. This time interval is an improper interval between the events "rocket ship passes first marker" and "rocket ship passes second marker." The interval is improper because the two events occur at two different places in this frame of reference

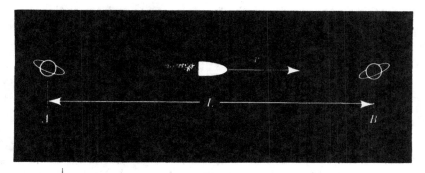

Figure 4-1

and, therefore, have to be timed with two different clocks. On the other hand, a clock on the rocket ship is present at both events and would, therefore, measure a proper time interval between them given by Equation 3-1 as

$$\text{proper time between passing the markers} = \frac{L}{v}\sqrt{1 - v^2/c^2} \quad (4\text{-}1)$$

In the frame of reference of the pilot, the markers approach him at the speed v. The first one passes by and then a time $(L/v)\sqrt{1 - v^2/c^2}$ later the second one passes. The pilot will find the distance L' between the two markers to be their (velocity) × (this time interval) or

$$L' = L\sqrt{1 - v^2/c^2} \quad (4\text{-}2)$$

In the frame of reference where the markers move at a velocity v, the distance between them is decreased by the factor $\sqrt{1 - v^2/c^2}$.

Although this demonstration is certainly the simplest, and shows most directly how the phenomenon of time dilation leads necessarily to that of length contraction, we can get further insight by considering it in two more ways.

4-2 THE SECOND ARM OF THE MICHELSON-MORLEY APPARATUS

In this section we derive the length contraction from a different point of view. We turn our attention to the idealized Michelson-Morley apparatus and consider the arm that moves parallel to its length. Again, in a frame of reference in which the apparatus is at rest, the light flash goes out and returns in a time $\Delta t = 2L/c$. In the frame of reference in which the apparatus moves to the right with a speed v, we will find it convenient to calculate the time $\Delta t_1'$ for the light to reach the mirror and the time $\Delta t_2'$ for it to return. From the results of the previous section we suspect that the arm may shrink, and we anticipate this result by calling its length L' in this second frame of reference. If the light takes a time $\Delta t_1'$ to reach the mirror, the light must not only travel the distance L', but also the

Length Measurements

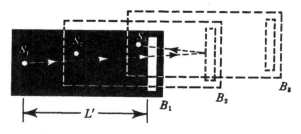

Figure 4-2

distance $v\Delta t_1'$ that the mirror has traveled. Thus

$$\Delta t_1' = \frac{L' + v\Delta t_1'}{c} \qquad \Delta t_1' = \frac{L'}{c - v}$$

We have made use of the fact that the light travels with speed c as required by the second postulate. The reader should show that the time for the return trip is

$$\Delta t_2' = \frac{L'}{c + v}$$

Finally, then

$$\Delta t' = \frac{L'}{c - v} + \frac{L'}{c + v} = \frac{\dfrac{2L'}{c}}{1 - \dfrac{v^2}{c^2}}$$

But $\Delta t'$ is an improper time interval between the sending and receipt of the light flash where $\Delta t = 2L/c$ is the proper time interval between the same two events. From Equation 3-1 we have

$$\Delta t = \Delta t' \sqrt{1 - \frac{v^2}{c^2}}$$

Then

$$\frac{2L}{c} = \frac{\frac{2L'}{c}}{1 - \frac{v^2}{c^2}} \sqrt{1 - \frac{v^2}{c^2}}$$

from which

$$L' = L\sqrt{1 - \frac{v^2}{c^2}} \tag{4-3}$$

as in the previous section. Here L is the length of the apparatus as measured by an observer at rest with respect to it, and L' is the length of the same apparatus measured by an observer moving with a speed v with respect to it. This last equation is sometimes summarized by the statement: Moving measuring rods shrink in the direction of their motion. Again it may be well to warn the reader against trying to find a mechanical cause for this "shrinkage." It is a kinematical effect related to the meaning of the length measurement.

4-3 A THIRD LENGTH MEASUREMENT

In the last two sections it was shown that an object that has a length L when measured in its own rest frame, appears to have a length $L' = L\sqrt{1 - v^2/c^2}$ when measured by an observer past which the object is moving. In this section we consider a gedanken experiment that is more directly related to how one would like to measure the length of a moving meter stick. Figure 4-3 shows a meter stick moving through the laboratory. This meter stick has a length L between the marks A and B when measured in its own rest frame, i.e., if A is the 12 cm mark and B the 87 cm mark, then $L = 75$ cm. As this meter stick passes through the laboratory, it is photographed beside another meter stick at rest in the laboratory. On the photograph it is found that points A and B are opposite the marks A' and B' on the laboratory meter stick. The photograph is made with a camera placed along the perpendicular bisector of $A'B'$. This is to ensure that the light left A and A' simultaneously with that which

Length Measurements

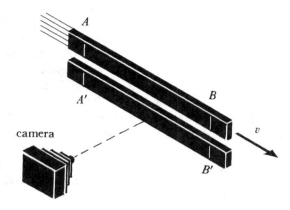

Figure 4-3

left B and B'. It would clearly be ridiculous to note the mark A' past which A was traveling and then wait a little while before noting the mark B' past which B was traveling. The meter stick would have moved between readings and anyone would protest that this method would be an incorrect way to make a length measurement. We suppose that we have taken a picture like that of Figure 4-4a. We find A beside A' and B beside B'. The experiment cannot actually be done, so the outcome must be calculated from previous results.

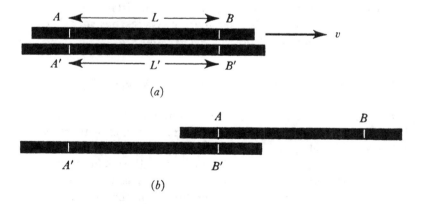

Figure 4-4

Figure 4-4a shows the configuration of the meter sticks when the picture was taken, and Figure 4-4b shows them at a later time when the mark A passes the mark B'. Consider the problem first in the frame of reference of the laboratory (the primed frame). In the laboratory the point A moved the distance L' from A' to B' at a speed v. Therefore it took a time L'/v. That is, clocks at rest in the laboratory advanced a time L'/v between the events "A passes A'" and "A passes B'." In particular a clock at B' advances L'/v during this time.

Now consider what happens from the point of view of an observer on the meter stick moving through the laboratory. From *his* point of view, the mark B' moved from B to A at a speed v. Since the distance from B to A is L, this took a time L/v. That is, clocks at A and B advanced L/v. But this time interval is an *improper time interval* between the events "B' passes B" and "B' passes A." The interval L'/v measured by the clock at B', present at both events, is a *proper time interval* between the same two events. These are related by Equation 3-1 which gives

$$\frac{L'}{v} = \frac{L}{v}\sqrt{1 - v^2/c^2}$$

or

$$L' = L\sqrt{1 - v^2/c^2} \tag{4-4}$$

Therefore, in the picture of Figure 4-4a the distance L fits against the shorter distance L'. The moving meter stick has shrunk by the same factor as that found in Sections 4-1 and 4-2.

4-4 THE LENGTH PARADOX AND SIMULTANEITY

Some readers will now recognize an apparent paradox. How can it be that the moving meter stick appears to shrink with respect to the one in the laboratory, when to an observer riding with the moving meter stick the laboratory is moving, and the laboratory meter stick must shrink with respect to it? It appears as if these statements are logical contradictions. The resolution of the paradox lies in another remarkable result of special relativity, namely, that two

Length Measurements

events which happen at the same time for one observer do not happen at the same time for a second observer moving relative to the first. In the case at hand, the events "A passes A'" and "B passes B'" were simultaneous in the laboratory and were, therefore, suited for comparing the lengths of the laboratory and "moving" meter sticks. To an observer on the "moving" meter stick the events do not occur simultaneously (as will be shown) and are, therefore, not suitable for a comparison of lengths.

The fact that simultaneous events are not simultaneous to a second moving observer means that careful thought must be given to the drawing and interpretation of diagrams in the framework of special relativity. A diagram, such as Figure 4-4a, purports to show an object moving past another *at one instant of time.* Insofar as this diagram shows a single instant of time, i.e., simultaneous events, it must be drawn from the point of view of a particular frame of reference. It may *not* be used, for example, to show the situation from the point of view of someone riding on the "moving" meter stick. Why not? In that frame of reference we shall show that A' does not pass A at the same time that B' passes B, and in that frame of reference, therefore, there is never a situation like that of Figure 4-4a.

In order to discuss the apparent paradox it seems that the meter sticks must be equipped with clocks. Figure 4-5a shows the clocks as A passes A' and B passes B'. The diagram is drawn in the laboratory frame of reference where the events are simultaneous. Therefore, the diagram shows the two clocks at A' and B' reading the same time, which has been taken to be zero arbitrarily. The clock on the "moving" meter stick at A reads some time T, but the reading of the clock at B has yet to be determined.

Next consider the later time when A passes B'. Figure 4-5b shows this situation. In this time interval A has traveled the distance L' from A' to B', and the laboratory clocks have therefore advanced to L'/v. This is an improper time interval between the events "A passes A'" and "A passes B'." The proper time interval is measured by the clock at A on the "moving" meter stick and is $(L'/v)\sqrt{1 - v^2/c^2}$. The clock at A, therefore, reads $T + (L'/v)\sqrt{1 - v^2/c^2}$ as shown. Remember there is nothing imaginary about this; Figure 4-5b could be a photograph. Anybody, even an observer in another refer-

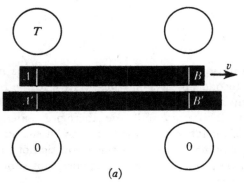

(a)

drawn in the primed frame

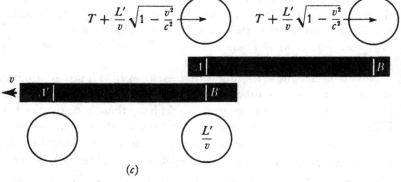

(b)

drawn in the primed frame

(c)

drawn in the unprimed frame

Figure 4-5. *(Continued on p. 71.)*

Length Measurements

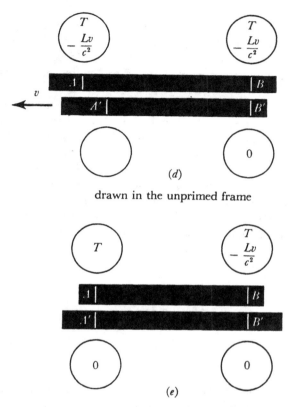

(d) drawn in the unprimed frame

(e) drawn in the primed frame

Figure 4-5

ence frame, could study that photograph. No matter what the reference frame, then, when A passes B', the clock at A reads $T + (L'/v)\sqrt{1 - v^2/c^2}$ and that at B' reads L'/v.

Consider again the instant when A passes B', but now from the frame of reference of the "moving" meter stick shown in Figure 4-5c. Again the clocks at A and B' read $T + (L'/v)\sqrt{1 - v^2/c^2}$ and L'/v as before, but now it is permissible to write $T + (L'/v)\sqrt{1 - v^2/c^2}$ for the clock reading at B. This is because the diagram shows an instant of time in the frame of reference of the "moving" meter stick, and all clocks in that frame of reference read alike. If Figure 4-5c were to be considered a photograph, it would have to be taken with a camera equidistant from A and B and riding along with the meter stick.

Finally consider Figure 4-5d, again in the frame of reference of the "moving" meter stick, at the previous instant when B' passed B. Notice that the diagram does not assume that A is passing A' at this instant; indeed, it is not. From Figure 4-5a we find that when B' passed B, the clock at B' read zero. Therefore, from the event "B' passes B" (shown in Figure 4-5d) to the event "B' passes A" (shown in Figure 4-5c) the clock at B' advances the proper time interval L'/v. The improper time interval between these events is measured by clocks at B and A; they therefore advance $(L'/v)/\sqrt{1 - v^2/c^2}$. The clock at B in Figure 4-5d therefore reads

$$T + \frac{L'}{v}\sqrt{1 - v^2/c^2} - \frac{(L'/v)}{\sqrt{1 - v^2/c^2}} =$$
$$T - \frac{L'}{v}\frac{v^2/c^2}{\sqrt{1 - v^2/c^2}} = T - \frac{Lv}{c^2} \quad (4\text{-}5)$$

But this is, of course, the reading of the clock at B when passing B' no matter in what frame of reference it is seen. We may, therefore, return to the original diagram or photograph as seen in the laboratory, Figure 4-5e.

We finally see the resolution of the "length paradox." In the laboratory the passage of A' past A and B' past B were simultaneous events, just what is necessary to compare the distance between A and B to that between A' and B'. In the frame of reference of the moving meter stick they are not simultaneous as evidenced by the fact that clocks at A' and B' do not read the same. In the frame of reference of the moving meter stick these events are not suitable for comparing a meter stick in that frame to one in the laboratory. It is left to the reader, in Exercise 2, to show that the difference in time is exactly right for both observers to find the same shrinkage factor for meter sticks moving through their frames of reference.

Length Measurements

4-5 LENGTHS PERPENDICULAR TO THEIR MOTION DO NOT CHANGE

In Chapter 3 we made the tacit assumption that the length of the arm of the Michelson-Morley apparatus moving perpendicular to itself was L in both frames of reference. Our results in the last sections should lead us to doubt this. However, our results were correct; lengths perpendicular to motion do not change as we shall show now.

Consider two measuring rods approaching each other at a speed v as shown in Figure 4-6. Each rod is equipped with markers that are the same distance apart when the rods are compared at rest. For purposes of argument, consider the frame of reference fixed with respect to rod B, i.e., rod A is moving toward B with a speed v. As rod A passes B it marks B permanently. Suppose rod A shrinks and, hence, marks B inside B's own markers. B must then have marked A outside its markers. This is a permanent record. Even an observer fixed on rod A, i.e., riding with it, can look back, see the marks, and must agree A is shorter. But to an observer fixed on rod A, B is the moving rod and appears *longer* than the fixed rod A. This situation violates the first postulate of relativity because it would provide a

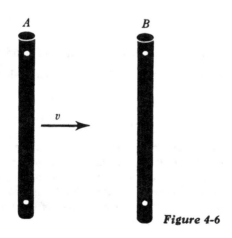

Figure 4-6

means for detecting a property of absolute motion, i.e., in our example, rods moving right shrink, rods moving left stretch. The only conclusion that fits all the postulates is that in either system neither rod changes length.

4-6 A SUMMARY

We have now obtained all the basic results of relativistic kinematics. It is therefore worthwhile to summarize them here:

1. If two events are seen that occur in the same place in one frame of reference, the time interval between them is called *proper* in that frame of reference and *improper* in a frame of reference moving with respect to the proper frame. The relation between the improper and proper time intervals is

$$\Delta t(\text{proper}) = \Delta t(\text{improper}) \sqrt{1 - v^2/c^2} \qquad (4\text{-}6)$$

or

Moving clocks run slow.

2. If two events occur a distance $\Delta L'$ apart in a frame of reference in which they occur simultaneously, then they are a distance ΔL apart in a frame of reference moving with a speed v in the direction in which ΔL is measured, and

$$\Delta L' = \Delta L \sqrt{1 - v^2/c^2} \qquad (4\text{-}7)$$

or

Moving meter sticks shrink in the direction of motion.

3. Meter sticks moving perpendicular to their lengths remain unchanged.

Exercises

1. The method used in Section 4-5 to show that meter sticks moving perpendicular to their lengths do not change length seems simple enough. It is clear that we cannot use quite the same method for meter sticks moving parallel to their length. However, a meter stick flying past the laboratory could have a device on it for lowering two pieces of chalk simultaneously, one meter apart, and marking the floor. The chalk marks would then either

Length Measurements

be farther apart or closer together than one meter in the laboratory. Why can we not go on with an argument just like that in Section 4-5 to obtain the result that meter sticks moving parallel to their length remain unchanged? Incidentally, will the chalk marks actually be closer together or farther apart than one meter?

2. Reference to Figure 4-5e shows that an observer on the "moving" meter stick at B did not wait long enough to make a measurement which was simultaneous to that of A passing A'. In fact, he needed to wait a time $Lv/c^2 = L'v/c^2 \sqrt{1 - v^2/c^2}$ longer. B would have been opposite a point C' when the B clock read T. Find the distance $B'C'$ (be careful about proper

and improper time intervals), and show that the distance $A'C'$ is longer than AB by the factor $\sqrt{1 - v^2/c^2}$. Thus, a properly made simultaneous measurement in the unprimed system also shows that moving meter sticks shrink.

3. A device is constructed which has a light flasher F midway between two mirrors, A and B, inclined at 45°. A and B are a distance L apart, and hence each is a distance $L/2$ from the flasher. The device moves through the laboratory at a speed v. The flasher flashes at time $t = 0$, as measured

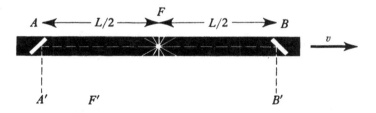

on the device and in the laboratory, just as it passes a point F' in the laboratory. The light flash travels to A and B and is reflected to the laboratory where it leaves a mark (perhaps photographically) at A' and B'. (The time taken to go from mirror to A' or B' is considered negligible.) Find the readings of clocks at A, B, A', and B' when the flash arrives, and find the distances A' and B' from the point F'. Is the distance between A' and B' what you expect? Did the light flash travel at a speed c in the laboratory as well as in the device's frame?

4. In Figure 4-5 several of the clocks were left blank. Why? If a camera had actually taken a photograph to make these figures, the clock would have shown some time on its face. Of course, these blank clocks would not have been equidistant with the other clocks from the camera, and the time of flight of the light would have to have been taken into account. But ignoring this effect, i.e., assuming simultaneous pictures of the clock faces in the appropriate frame of reference, calculate the readings on all the blank clocks.

5. A rocket ship of length L (in the rocket's frame) leaves the earth at a speed $(\frac{4}{5})c$. A light signal is sent after it which arrives at the rocket's tail at $t = 0$ according to rocket clocks and earth clocks.

a. When does the signal reach the head of the rocket according to rocket clocks?

b. According to earth clocks?

c. The answers to parts a and b are not related like proper and improper time intervals. Why?

The light signal is reflected from the head back to the tail.

d. When does it reach the tail according to rocket clocks?

e. According to earth clocks?

f. Are the answers to d and e related like proper and improper time intervals? Why?

5

Velocity and Acceleration

5-1 ADDING VELOCITIES

IT OFTEN HAPPENS THAT THE VELOCITY of an object is known in one frame of reference, which in turn is moving with respect to a second frame of reference, and we wish to find the velocity of the object in this second frame of reference. A familiar example is that of an airplane pilot who finds himself flying at 550 mi/hr due north with respect to the air. There is, however, a cross wind blowing 150 mi/hr east with respect to the ground. The velocity of the plane with respect to the ground is the vector sum of these two velocities. See Figure 5-1.

The relativistic treatment of this same problem is not quite so simple, but we start just as we would classically. Figure 5-2a shows an object, a rocket ship, at the origin of a coordinate system O. Figure 5-2b shows the object at a later time T where it has moved to a position $x = X, y = Y$. The time interval between these two events, the start at O and the arrival at (X,Y), is measured by clocks at rest in O. During this same time a second coordinate system O' has moved a distance vT out along the x-axis, also as shown in Figure 5-2b. It may be convenient to think of the coordinate systems O and O' as actual frameworks with respect to which measurements are

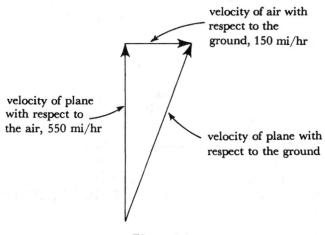

Figure 5-1

made. The velocity U of the object in O is defined to be the distance moved in O divided by the time measured in O. Therefore, we have

$$U_x = \frac{X}{T} \qquad U_y = \frac{Y}{T} \tag{5-1}$$

An observer riding on O' views this trip as seen in Figure 5-2c and 5-2d. The object starts at O' and arrives at the end of its trip at $x' = X'$ and $y' = Y'$ when clocks in O' read a time $t' = T'$. During this time interval the origin O of the first coordinate system has moved to the left a distance vT'. Accordingly,

$$U_x' = \frac{X'}{T'} \qquad U_y' = \frac{Y'}{T'} \tag{5-2}$$

Notice that Figures 5-2a and 5-2c as well as 5-2b and 5-2d look much the same. They are drawn separately to emphasize the fact that you must use one frame of reference at a time to do a kinematic calculation.

For example, the reader may say that S and X' are the same. After all, are they not both the distance from O' to the end of the trip? X' is the distance from O' to the point marking the end of the trip. It may be considered an actual distance marked off on the O' frame-

Velocity and Acceleration 79

work. S is this same distance interval, but seen by an observer in O past whom the marked-off distance is moving. In other words, Figure 5-2b is drawn at the instant $t = T$ everywhere in the O system. From Equation 4-7, $S = X'\sqrt{1 - v^2/c^2}$. Similarly, the distances vT and vT', although they are both the distance from O to O' simultaneously with the end of the trip, are not the same, because

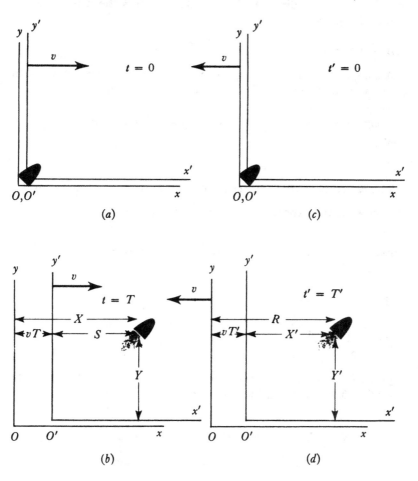

Figure 5-2. (a) and (b) are to be thought of as drawn from the point of view of an observer in O; (c) and (d) from the point of view of O'.

"simultaneously" means something different to the two different observers.

However, as long as we stay in one particular frame of reference, we may do calculations in the familiar classical way. For example, we look at Figure 5-2b and write $X = vT + S$. The important thing to notice here is that *all* quantities are written as seen in the O frame of reference—including times! Substituting for S the contracted X' found above gives

$$X = vT + S = vT + X'\sqrt{1 - v^2/c^2} \tag{5-3}$$

A completely equivalent expression can be derived by starting with the O' system. The reader is asked to complete the following sentences (answers are at end of the chapter).

Looking at Figure 5-2d, and using quantities completely in the O' frame,

$X' = $ _____ (a).

But R can be written in terms of X as $R = $ _____ (b).

This gives the expression equivalent to Equation 5-3 $X' = $ _____ (c).

Equation 5-3 and its equivalent, (c), can be solved for X' and T' in terms of X and T

$$X' = \frac{X - vT}{\sqrt{1 - v^2/c^2}}$$

$$T' = \frac{T - vX/c^2}{\sqrt{1 - v^2/c^2}} \tag{5-4}$$

Equation 5-4 immediately gives the desired result for U_x. Substituting Equation 5-4 into Equation 5-2 gives

$$U_x' = \frac{X'}{T'} = \frac{X - vT}{T - \frac{vX}{c^2}} = \frac{\frac{X}{T} - v}{1 - \frac{vX/T}{c^2}} = \frac{U_x - v}{1 - \frac{vU_x}{c^2}}$$

The y-coordinate of the end of the trip is a distance measured

Velocity and Acceleration

perpendicular to its direction of motion so that $Y = Y'$. Thus

$$U_y' = \frac{Y'}{T'} = \frac{Y}{T - \frac{vX}{c^2}} = \frac{U_y\sqrt{1 - v^2/c^2}}{1 - \frac{vU_x}{c^2}}$$

Summarizing, then

$$U_x' = \frac{U_x - v}{1 - \frac{vU_x}{c^2}} \qquad U_y' = \frac{U_y\sqrt{1 - v^2/c^2}}{1 - \frac{vU_x}{c^2}} \qquad (5\text{-}5)$$

These are the equations that relate velocities measured in two different frames of reference. They are often referred to as the *relativistic velocity addition equations*.

5-2 EMISSION OF LIGHT BY MOVING OBJECTS

If an atom, moving through the laboratory at a speed v along the x-axis, emits light at an angle θ to the x-axis, as measured in its own frame of reference, the light is pushed forward in the laboratory frame of reference to a new angle θ'. Figure 5-3a shows the atom at rest in its own frame of reference O emitting light at an angle θ. Figure 5-3b shows it moving through the laboratory system O' where

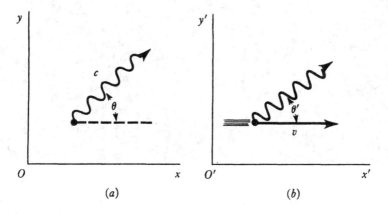

Figure 5-3

the new angle of emission is θ'. With respect to the atom the light travels at a speed c, and therefore has velocity components

$$U_x = c \cos \theta \qquad U_y = c \sin \theta$$

The laboratory (the primed system) moves at a speed $-v$ with respect to the atom, and Equation 5-5 gives, for the velocity components of the light with respect to the laboratory,

$$U_x' = \frac{c \cos \theta - (-v)}{1 - \frac{(-v)c \cos \theta}{c^2}} = \frac{c \cos \theta + v}{1 + \frac{v \cos \theta}{c}}$$

$$U_y' = \frac{c \sin \theta \sqrt{1 - v^2/c^2}}{1 + \frac{v \cos \theta}{c}}$$

(5-6)

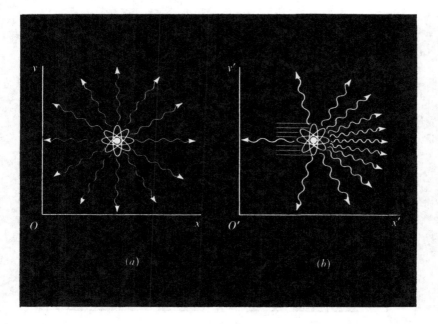

Figure 5-4. (a) *Suppose that an atom at rest emits light uniformly in all directions. (b) Then, when the atom is viewed from a frame of reference where it is moving with a speed 0.8c, for example, the light is thrown predominantly in a forward direction. Is the speed of the light changed? Is its frequency changed?*

Velocity and Acceleration

The angle θ' at which the light travels in the laboratory is given by

$$\tan \theta' = \frac{U_y'}{U_x'} = \frac{\sin \theta \sqrt{1 - v^2/c^2}}{\cos \theta + v/c} \quad \text{or}$$

$$\sin \theta' = \frac{\sin \theta \sqrt{1 - v^2/c^2}}{1 + \frac{v \cos \theta}{c}} \quad (5\text{-}7)$$

If the moving object emits light uniformly in all directions, just half of it goes forward of $\theta = 90°$ or $\sin \theta' = \sqrt{1 - v^2/c^2}$. For objects traveling with nearly the speed of light, θ' becomes very small. Rapidly moving objects therefore emit light in predominantly forward directions. Figure 5-4 shows what happens for an atom moving at a speed $0.8c$.

5-3 ACCELERATION

Suppose a spaceship accelerates with a constant acceleration equal to that of gravity on the earth's surface. How long will it take to go to the nearest star? So far, all our dealings have been with uniformly moving objects and systems. Since this question involves accelerations, one might wonder whether special relativity can supply an answer at all. The answer is that it can, so long as we ask questions about where the spaceship goes and how long it takes *as measured in an inertial system*. On the other hand, we may not ask questions about things seen from the point of view of travelers on the ship, for that is not an inertial system.

First of all we must know what is meant by a constant acceleration. We mean an acceleration that is constant for those on board the spaceship. In other words, enough fuel is ejected so that the ship is accelerated with an acceleration a_0 as measured by accelerometers on board the spaceship. The point is, that, unlike a uniform velocity, an acceleration is something that can be measured inside a completely closed laboratory.

Suppose we wish to describe the accelerated motion of the spaceship in an inertial frame of reference O. No matter how complicated the motion, the spaceship has some velocity v at each instant of time. This discussion will be concerned with motion in the x-direction only. Since the motion is accelerated, the spaceship will have a velocity $v + \Delta v$ at a time Δt later. All these quantities are to be measured in the frame of reference O. If the motion of the spaceship is described from a second inertial reference frame O', which is moving with respect to O with the velocity v, the spaceship will appear momentarily at rest, Figure 5-5a. As long as velocities remain small in this new frame of reference, the motion may be treated classically, since Newtonian mechanics is valid for small speeds. For instance, if accelerometers aboard the spaceship measure an acceleration a_0, then the acceleration of the ship will be a_0 with respect to the new reference frame. The spaceship will not stay at rest, but will accelerate to a velocity $\Delta v' = a_0 \Delta t'$ after a time $\Delta t'$ measured in this new frame of reference, O', Figure 5-5b. Notice that it is at this point in the argument, where we may apply classical concepts, that contact is made between what goes on in the spaceship and what goes on in an inertial frame.

When the velocity of the ship is zero in O' it is, of course, v in the original system O. When the velocity of the ship has reached $\Delta v'$ in O', its velocity in the system O is given by the relativistic velocity addition formula Equation 5-5

$$v + \Delta v = \frac{v + \Delta v'}{1 + \frac{v \Delta v'}{c^2}} \tag{5-8}$$

Since $\Delta v' \ll c$, even though v may be comparable to c, Appendix A may be used to expand the denominator

$$\left(1 + \frac{v \Delta v'}{c^2}\right)^{-1} \approx \left(1 - \frac{v \Delta v'}{c^2} + \cdots\right)$$

Then Equation 5-8 becomes

$$v + \Delta v = (v + \Delta v')\left(1 - \frac{v \Delta v'}{c^2}\right) = v + \Delta v'\left(1 - \frac{v^2}{c^2}\right)$$

Velocity and Acceleration

where all terms in $(\Delta v')^2$ have been neglected. The change in velocity, as measured in O, is therefore

$$\Delta v = v + \Delta v'(1 - v^2/c^2) - v = \Delta v'(1 - v^2/c^2) \tag{5-9}$$

Now this change in velocity occurred over a time $\Delta t'$ in O' and Δt in O. In the system O' all velocities are small provided $\Delta t'$ is kept small.

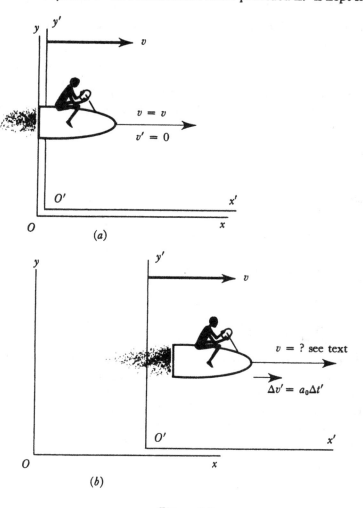

Figure 5-5

The events "ship is at the origin with velocity zero" and "ship is somewhere with velocity $\Delta v'$" occur essentially at the same place in O'. The time interval $\Delta t'$ is, therefore, a proper time between them. Δt and $\Delta t'$ are therefore related as

$$\Delta t' = \Delta t \sqrt{1 - v^2/c^2} \tag{5-10}$$

Combining Equations 5-9 and 5-10 and remembering that $\Delta v' = a_0 \Delta t'$, we obtain

$$a = \Delta v/\Delta t = a_0(1 - v^2/c^2)^{3/2} \tag{5-11}$$

where the acceleration a has been written for $\Delta v/\Delta t$.

To find the velocity as a function of time, Equation 5-11 must be integrated. Thus

$$\frac{dv}{dt} = a_0(1 - v^2/c^2)^{3/2}$$

$$a_0 dt = \frac{dv}{(1 - v^2/c^2)^{3/2}}$$

$$a_0 t = \frac{v}{(1 - v^2/c^2)^{1/2}} + \text{constant}$$

If $v = 0$ at $t = 0$, the constant is zero and

$$v = \frac{c}{\sqrt{1 + \frac{c^2}{a_0^2 t^2}}} = \frac{a_0 t}{\sqrt{1 + \frac{a_0^2 t^2}{c^2}}} \tag{5-12}$$

The first expression is convenient for large t, the second for small t. For very small t, $v = a_0 t$, as it should, since if t is small enough, no velocities can get large, and the relativistic result must equal the classical result. For very large t, the velocity approaches c.

Integration of Equation 5-12 gives the distance traveled as a function of time.

$$x = \frac{c^2}{a_0}\sqrt{1 + \frac{a_0^2 t^2}{c^2}} - \frac{c^2}{a_0}$$

where $x = 0$ at $t = 0$.

Velocity and Acceleration

5-4 THE ULTIMATE SPEED

Suppose a spaceship is cruising past earth at a speed $.99c$ and shoots forward a smaller projectile at a speed $.99c$ with respect to it. Equation 5-5 tells us very quickly that the projectile is going at a speed

$$U_x = c\,\frac{.99 + .99}{1 + .99(.99)} = .99995c$$

with respect to the earth.

Suppose a spaceship accelerates with constant acceleration (as measured by its passengers) forever. Equation 5-12 tells us it comes up to the speed c, but never surpasses it. It looks very much as if there is an ultimate speed that cannot be exceeded, and that speed is the speed of light. The prediction of this ultimate speed is one which is characteristic of relativity.

Although this result has been believed on the basis of much experience over a great number of years, a very recent experiment has been performed by W. Bertozzi which is very appealing because

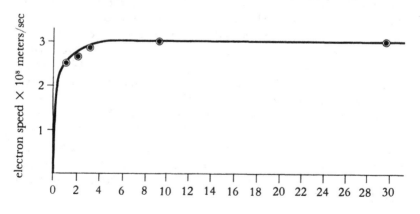

Figure 5-6

of its directness and basic simplicity.[1] Bunches of electrons were formed in a machine called a linear accelerator. These electrons were accelerated to high speeds that were directly determined by timing the flight of the electrons over a measured distance. The *number* of electrons was determined by measuring the total electric charge in the beam; the total energy in the beam was measured very directly by its heating effects in a calorimeter. The energy per electron was then computed. Figure 5-6 shows a graph of the speed of the electrons as a function of their energy. We will discuss the exact shape of the curve in a later chapter. Here we need be concerned with only one aspect of it. As the energy of the electrons gets very great, their speed seems to approach, but never exceed, the speed of light.

5-5 THE DOPPLER EFFECT

In the case of ordinary periodic waves in a mechanical medium, motion of either the source or the observer through the medium alters the observed frequency. This phenomenon, the Doppler effect, has been discussed in Chapter 2. A similar effect is observed in the case of light. In the latter case, according to the theory of relativity, only the relative velocity between source and observer can affect the observed frequency. In order to discuss the effect, we are at liberty to pick any convenient frame of reference, and we choose the one where the source is moving and the observer is stationary. We proceed exactly as in Chapter 2. Figure 5-7 shows the time when the observer O has just received a signal from S_1. We call $t = 0$ the time at which the signal left S_1. It will arrive at O a time r_1/c later. The source will emit a second signal or wave when it has arrived at S_2 at a time $t = \Delta t_s$, and this wave will arrive at O at a time $t = \Delta t_s + r_2/c$. The time between receipt of two successive signals at O will be $\Delta t_o = \Delta t_s + r_2/c - r_1/c$. If the source is very far from the observer, it is clear that

$$r_1 - r_2 = \overline{S_1 S_2} \cos \theta \quad \text{(see page 24)}$$

[1] W. Bertozzi, *Am. J. Phys.* **32**, 551 (1964).

Velocity and Acceleration

and that

$$\Delta t_o = \Delta t_s - \frac{\overline{S_1 S_2}}{c} \cos \theta$$

But $\overline{S_1 S_2} = v \Delta t_s$ since v is the speed with which the source is moving, and

$$\Delta t_o = \Delta t_s \left(1 - \frac{v}{c} \cos \theta \right)$$

So far everything is *exactly* as it would be classically for an observer at rest in the medium. The reason is that, so far, all distances and times have been measured in the frame of reference of the observer. We now calculate Δt_s in terms of the frequency of the source f_s. $1/f_s$ is the time between emission of two signals or waves by the source, as measured at the source. It is therefore a proper time interval. Δt_s is the time interval between the same two events, but in the frame of reference of the observer. These time intervals are related like proper and improper times, i.e.,

$$1/f_s = \Delta t_s \sqrt{1 - v^2/c^2}$$

We have

$$\Delta t_o = \frac{1 - \frac{v}{c} \cos \theta}{f_s \sqrt{1 - v^2/c^2}}$$

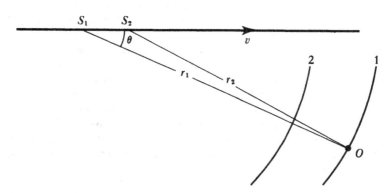

Figure 5-7

Since the frequency observed by the observer is just $1/\Delta t_o$ we have

$$f_o = f_s \frac{\sqrt{1 - v^2/c^2}}{1 - \frac{v}{c}\cos\theta} \tag{5-13}$$

For the particular case of $\theta = 0$ (source directly approaching the observer)

$$f_o = f_s \sqrt{\frac{1 + v/c}{1 - v/c}} \tag{5-14}$$

A non-relativistic derivation of Equation 5-13 gives the same denominator, whereas the numerator is directly related to relativistic time dilation. The experimental observation of this term by Ives and Stilwell in the frequency of light emitted by moving atoms (see page 57) was the first experimental verification of relativistic time dilation.

Exercises

1. A rocket moves past the earth at a speed $\frac{3}{5}c$, and at a later time sends back a "scout" rocket which travels at a speed $\frac{4}{5}c$ with respect to the "parent" rocket. With what speed does the "scout" rocket travel with respect to the earth?

2. A K^+ meson at rest decays into two π mesons which each travel with speed of about $.85c$. If a K^+ meson is traveling at a speed $.9c$ through the laboratory when it decays, what is the greatest and what is the least speed one of the π mesons could have?

3. Suppose a star is seen at right angles to the earth's velocity about the sun. By how much is the star's light deviated because of aberration? Is the relativistic result the same as the classical? By how much do they differ?

4. A π^0 meson decays into two photons which travel off in opposite directions in the frame of reference of the π^0 meson. You may consider the two photons to be flashes of light. If the meson is traveling through the laboratory with a speed $.9c$ when it decays, both photons going at right angles to its direction of travel as measured in the π^0 rest frame, what angle do the photons make with the direction of the π^0 meson in the laboratory?

5. Two rocket ships leave the earth at equal speeds of $c/2$ and make equal angles of $60°$ with the x-direction. Find the relative velocity of the two rocket ships.

Velocity and Acceleration

6. Solve Equations 5-5 for U_x and U_y in terms of U_x' and U_y'. How do you know your answer is correct?

7. Show that the velocity addition formulas are consistent with the second postulate by showing that an object with speed c in one frame of reference has a speed c in any other, even though the direction of travel may be different in the two frames.

8. In Section 5-1 we discussed two times T and T' for the duration of a trip from the origin to (X,Y). Neither of these times is proper for the trip. In what frame of reference is the proper time measured? What is this proper time in terms of T and U_x and U_y? In terms of T', U_x', and U_y'?

9. If a spaceship accelerated constantly with the acceleration of gravity at the earth's surface, i.e., in its own reference frame, how long would it take to go from here to α-Centauri, four light-years distant? Time is to be measured by earth clocks.

10. Recompute the Doppler effect in the frame of reference where the *source* is at rest. Remember, the observed frequency is that observed in the observer's own reference frame. Is your result the same as Equation 5-13? In comparing, remember θ may be different because of aberration.

11. There is a way to derive the velocity addition formula directly from the Doppler effect. Consider a frame of reference where *both* the source and the observer are moving. The source moves at speed U_s and the observer at speed U_o. In this reference frame compute the observed frequency f_o in terms of the source frequency f_s and the velocities U_s and U_o. However, relativity demands that the result be given in terms of the relative velocity only, as in Equation 5-14. What must this relative velocity be in terms of U_s and U_o in order that your answer be consistent with Equation 5-14? Do the collinear case only.

12. A material is said to have a refractive index n if the speed of light in the medium is c/n. If a piece of transparent material of the index of refraction n is moving through the laboratory at a speed v, and a flash of light is moving through the medium in the direction of its motion, what is the speed of the flash of light with respect to the laboratory? Show that this speed is intermediate between c/n and $c/n + v$. This "dragging" of the light along with the medium was measured before relativity was known, and it was interpreted as a partial dragging of the ether by a moving material medium, called the Fresnel drag.

13. A spaceship A sets out from the earth with a speed v at a time when its clocks and earth's read zero. A time T later, by the earth's clocks, a second spaceship B sets out at a speed u. Since u is presumed greater than v, the second spaceship catches the first at a time t, such that $(t - T)u = tv$, or $t = uT/(u - v)$.

a. When, according to clocks on ship A, did ship B leave earth?
b. When, according to clocks on ship A, did B catch up to A?
c. In A's reference frame, how far away was the earth when B started?
d. From (*a*), (*b*), and (*c*) deduce the velocity of ship B as measured by ship A. Does it agree with Equation 5-5?

14.
An observer O sees light coming to him at 30° to the direction of travel of an object A which is passing through the laboratory at a speed .99c. At what angle did the light leave A as measured in the rest frame of A? Does O see the back of A as A approaches?

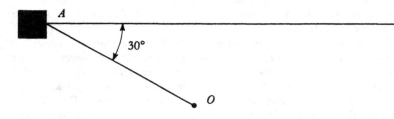

Answers to completion sentences on page 80

a. $X' = R - vT'$
b. $R = X\sqrt{1 - v^2/c^2}$
c. $X' = X\sqrt{1 - v^2/c^2} - vT'$

6

The Twin Paradox

6-1 STATEMENT OF THE PARADOX

AN OBSERVER ON EARTH watching a clock recede rapidly into space concludes that the receding clock runs slower than clocks fixed on the earth. An observer riding with that clock into space, however, concludes that it is the clocks on earth that are running slow. As stated in these terms, the situation sounds completely paradoxical. The paradox disappears, as we saw in Chapter 3, when we recall that those judgements were made on the basis of information on different sets of events. The situation is strange to our classically conditioned sense of time, but there is no logical inconsistency.

A somewhat different situation arises in the so-called "twin paradox." Two twins are at rest on the earth. One takes a very rapid rocket ride to a neighboring planet. While he is on this trip, the earth twin sees the rocket twin's clocks running slow. Among all possible clocks are biological processes; so, in fact, the earth twin thinks the rocket twin is aging more slowly than he himself is. The same thing is true of the return trip, since time dilation depends only on the square of the velocity. At the end of the trip, therefore, the two twins are standing side by side, but the rocket twin is younger than the earth twin. This is a startling conclusion, but most physicists believe that *this is the correct conclusion* from relativity. The *paradox*

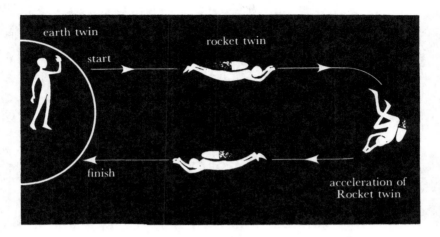

Figure 6-1

arises when we ask what the rocket twin thinks. *He* sees the earth twin aging more slowly then he, and when he returns to earth, he thinks the earth twin is the younger. We will subsequently show that this conclusion has faulty reasoning in it and is incorrect.

However, are not the two situations identical? How did the rocket twin know that the earth twin did not go out on a rocketship (the earth) and return? The physical difference is that the rocket twin *accelerated* at the end of the trip and the earth twin did not, Figure 6-1. Acceleration is something that can be observed physically, and hence the acceleration becomes an *event*. The trip then consists of *three distinct events*. The *start*, the *finish*, and the *acceleration of the rocket twin*. The earth twin is present at only two of the events, the rocket twin at all three. The trip, then, is not symmetrical between the two twins, and it is allowable for both twins to conclude that the rocket twin is the younger.

The resolution of the paradox is seen to involve a discussion of accelerations of two different observers. Properly, the general theory of relativity must be invoked to talk about such accelerations, but that complete a discussion is beyond the scope of this book. However, a rather satisfying discussion of the problem can be made in the framework of special relativity alone, and this detailed discussion follows.

The Twin Paradox

6-2 THE SOLUTION IN TERMS OF TIME DILATION

For convenience we modify the method for making the trip (Figure 6-2) so that there are two rocketships, one always traveling outward from the earth at a speed v, the other always headed in toward the earth at a speed v. The rocket twin makes his trip by jumping on the outbound ship, changing ships at the end of his trip, and jumping off as the second ship passes the earth. We now have three observers always in inertial systems, the earth twin and the two rocketship *pilots*.

Let us look at the trip from the point of view of the *earth twin*. If the whole trip takes a time T, the outbound trip takes a time $T/2$, and the inbound trip takes a time $T/2$. These times are entered in Table 6-1. Next, let us look at the trip from the point of view of the *outbound pilot*. The time for the outbound trip is a proper time for the outbound pilot, since the start event and change event occur at the

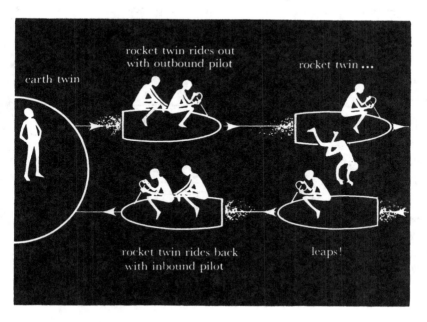

Figure 6-2

same place in his frame of reference, namely, his rocketship. The time interval between these same events is improper for the earth twin, and we know the ratio between these time intervals. The pilot sees a time $(T/2)\sqrt{1 - v^2/c^2}$ for the time interval of the outbound trip. This is entered in the first column of Table 6-1. The time interval for the whole trip is proper for the earth twin since he is present at both start and finish. It is improper for the outbound pilot. Again we know the ratio between the two observed intervals and find that the whole trip appears to take a time $T/\sqrt{1 - v^2/c^2}$ to the outbound pilot. This is entered in the last column of the table. To this pilot, then, the inbound trip takes a time $T/\sqrt{1 - v^2/c^2} - (T/2)\sqrt{1 - v^2/c^2}$, namely the whole time minus the time for the outbound trip.

Similar reasoning gives the times for the inbound and outbound trips as seen by the inbound pilot. These are also recorded in Table 6-1.

Now, in order to avoid confusion let us assume that the rocket twin carries *no watch*, but calculates the time of his trip by asking the pilots of his rocketships. He hops on the first, and just before he gets off he asks the outbound pilot how long the trip took. His answer: $(T/2)\sqrt{1 - v^2/c^2}$ from the outbound pilot's record of the trip out,

Table 6-1. *Trip Time*

	Outbound	Inbound	Whole
As seen by earth twin	$T/2$	$T/2$	T
As seen by outbound pilot	$\dfrac{T}{2}\sqrt{1 - v^2/c^2}$	$\dfrac{T}{\sqrt{1 - v^2/c^2}} - \dfrac{T\sqrt{1 - v^2/c^2}}{2}$	$T/\sqrt{1 - v^2/c^2}$
As seen by inbound pilot	$\dfrac{T}{\sqrt{1 - v^2/c^2}} - \dfrac{T\sqrt{1 - v^2/c^2}}{2}$	$\dfrac{T}{2}\sqrt{1 - v^2/c^2}$	$T/\sqrt{1 - v^2/c^2}$

The Twin Paradox

Table 6-1. Similarly, after the trip in, he asks the pilot how long he has been on board. Answer: $(T/2)\sqrt{1-v^2/c^2}$. He concludes the entire trip took $T\sqrt{1-v^2/c^2}$. The earth twin thinks the trip took a time T, but at the same time he has watched the rocket twin's clocks always running slow by the factor $\sqrt{1-v^2/c^2}$. The earth twin therefore agrees that the rocket twin should have measured the time of the trip to be $T\sqrt{1-v^2/c^2}$. Thus, both agree on all points.

6-3 THE SOLUTION IN TERMS OF "HEARTBEATS"

Lest some wary readers worry about the previous solution in terms of biological processes, let us have each twin count the other's heartbeats. We require each twin to wear a device that flashes a light on top of his head every time his heart beats. Furthermore, we will assume that each twin's heart beats once every second.

Let us start with the earth twin. He thinks the whole trip took T seconds, and therefore he knows his own light flashed and heart beat T times during the trip. What does he see happening to the rocket twin? The rocket twin's light is flashing with one flash per second. The earth twin sees these flashes Doppler shifted and actually sees $\sqrt{1-v^2/c^2}/(1+v/c)$ flashes per second on the outward trip and $\sqrt{1-v^2/c^2}/(1-v/c)$ flashes per second on the inward trip (Equation 5-13). He does *not*, however, see these frequencies for a time $T/2$ each. He sees the outbound frequency for a time $T/2$ *plus* the time it takes the light to get back to him from the turn-around point. Since the earth twin knows the outbound trip took $T/2$ seconds at a velocity v, the turn-around point is a distance $vT/2$ from the earth. The time it takes light to go this distance is $vT/2c$. He then sees the frequency $\sqrt{1-v^2/c^2}/(1+v/c)$ for a time $T/2 + vT/2c$ and the frequency $\sqrt{1-v^2/c^2}/(1-v/c)$ for a time $T/2 - vT/2c$. Multiplication shows that he sees $T\sqrt{1-v^2/c^2}$ flashes from the rocket twin. Therefore, the earth twin says: "my heart beat T times, rocket twin's beat $T\sqrt{1-v^2/c^2}$ times."

Now let us ask what the rocket twin sees. By his clocks the whole trip took a time $T\sqrt{1-v^2/c^2}$ so he agrees that his heart beat

$T\sqrt{1 - v^2/c^2}$ times. On the trip out he sees a Doppler shifted frequency of the earth twin's light $\sqrt{1 - v^2/c^2}/(1 + v/c)$ for a time $(T/2)\sqrt{1 - v^2/c^2}$ or $(T/2)(1 - v^2/c^2)/(1 + v/c)$ heartbeats. On the trip back he sees a frequency $\sqrt{1 - v^2/c^2}/(1 - v/c)$ for a time $(T/2)\sqrt{1 - v^2/c^2}$ or $(T/2)(1 - v^2/c^2)/(1 - v/c)$ heartbeats. He sees

$$\frac{T}{2}\left(\frac{1 - v^2/c^2}{1 + v/c} + \frac{1 - v^2/c^2}{1 - v/c}\right) = \frac{T}{2}(1 - v/c + 1 + v/c) = T$$

heartbeats from earth twin. Now we see the point where the sloppy reasoning that led to the paradox breaks down. The rocket twin indeed sees earth twin's clocks running slow. That is what gives rise to the $\sqrt{1 - v^2/c^2}$ term in the Doppler formula. However, if we count carefully how many heartbeats he sees from earth twin we see there is no paradox and he agrees: "Earth twin's heart beat T times, mine beat $T\sqrt{1 - v^2/c^2}$ times."

6-4 SOLUTION IN TERMS OF "HEARTBEATS" COUNTED BY THE OUTBOUND PILOT

As a final check, but more as a good exercise in relativity, let us count the heartbeats as seen by the outbound pilot. The pilot knows when the rocket twin jumped on, when he jumped off, and when he receives a light signal indicating the end of the trip. If he wants to know how many times the earth twin's heart beat during the trip, he will simply count heartbeat flashes from the earth twin from the start of the trip until the light reaches him indicating the end of the trip. Let the time from the start of the trip to receipt of this signal be $\mathfrak{J}$ as measured by the rocket ship's clocks. Then $\mathfrak{J}$ = (time until end of trip) + (time for light signal to come). The end of the trip was at $T/\sqrt{1 - v^2/c^2}$ and the light signal then had to go a distance $vT/\sqrt{1 - v^2/c^2}$. This takes a time $(vT/c)/\sqrt{1 - v^2/c^2}$. Thus

$$\mathfrak{J} = \frac{T}{\sqrt{1 - v^2/c^2}} + \frac{vT/c}{\sqrt{1 - v^2/c^2}} = \frac{1 + v/c}{\sqrt{1 - v^2/c^2}} T$$

The Twin Paradox

During this time he is seeing flashes at a frequency $\sqrt{1 - v^2/c^2}/(1 + v/c)$. The total number of flashes seen is frequency × time = T. (Note: This is not really a proof of anything. It is, in fact, an alternate way for deriving the Doppler effect.)

Now what does he see from the rocket twin? On the trip out he sees $(T/2)\sqrt{1 - v^2/c^2}$ flashes, the same as the rocket twin himself sees. The trip back is more complicated. The time from the start of the whole trip until he *sees* the end of the whole trip is

$$\mathfrak{I} = \frac{T(1 + v/c)}{\sqrt{1 - v^2/c^2}}$$

as before. Since the twin jumps off at a time $(T/2)\sqrt{1 - v^2/c^2}$ after the start, the total time he sees Doppler shifted flashes from the return trip is

$$\begin{aligned}\text{(Time flashes seen from return trip)} &= \frac{(1 + v/c)T}{\sqrt{1 - v^2/c^2}} - \frac{T}{2}\sqrt{1 - v^2/c^2} \\ &= \frac{1 + v/c}{1 - v/c}\sqrt{1 - v^2/c^2}\,\frac{T}{2}\end{aligned}$$

As far as the outbound pilot is concerned, the rocket twin is receding from him with a velocity U given by the velocity addition formula, Equation 5-5.

$$U = \frac{v + v}{1 + v^2/c^2} = \frac{2v}{1 + v^2/c^2}$$

The frequency he observes is given by the Doppler formula as

$$\frac{\sqrt{1 - \left(\frac{2v/c}{1 + v^2/c^2}\right)^2}}{1 + \frac{2v/c}{1 + v^2/c^2}} = \frac{1 - v^2/c^2}{(1 + v/c)^2} = \frac{1 - v/c}{1 + v/c}$$

The number of flashes is again given by the frequency × time or

$$\frac{1 - v/c}{1 + v/c} \times \frac{1 + v/c}{1 - v/c}\sqrt{1 - v^2/c^2}\left(\frac{T}{2}\right) = \sqrt{1 - v^2/c^2}\,\frac{T}{2}$$

The total number of flashes seen from the rocket twin on the outbound and inbound trips is thus $T\sqrt{1 - v^2/c^2}$. So finally even the pilot of the outbound rocketship sees T flashes from earth twin and $T\sqrt{1 - v^2/c^2}$ from the rocket twin. The times and frequencies seen by the outbound pilot are shown diagrammatically in Figure 6-3.

6-5 DISCUSSION AND EXPERIMENT

When we noted that there were really three physically observable events which make up the twin paradox problem, and that the experience of the two twins was not symmetrical, the logical paradox vanished. It only remained to see whether special relativity could provide an unambiguous solution to this problem which involved an acceleration. In our treatment, the acceleration experienced by the traveling twin was evidenced by the necessity for using two inertial systems to describe the problem from his point of view. We found that our theory was adequate for a complete description.

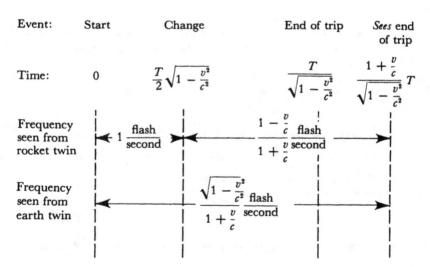

Figure 6-3. *This chart shows the times of the events and light flash frequencies seen between them, according to the outbound pilot.*

The Twin Paradox

The argumentative reader will observe, and quite properly, that we have really only sidestepped the acceleration question. What would have happened in the tremendous acceleration at turnabout if the rocket twin had carried clocks? We could take one more step with special relativity, treating a gradual turnabout as we did the accelerating object in Chapter 5; the result agrees with the one given here. However, since we are really asking for an observation made in an accelerated reference frame, we should be satisfied only by a treatment with general relativity. Suffice it to say that such a treatment agrees with ours provided the mass of the traveling rocket ship is small compared to the mass of the system in which it accelerates, i.e., the universe!

Most satisfying of all would be an appeal to experiment. Fortunately, we can do just that. Certain radioactive nuclei emit γ-rays (high energy electromagnetic waves just like light) whose frequency can be measured, in certain circumstances, to fantastic precision. Changes in the frequency of this radiation smaller than one part in 10^{13} can be detected. For instance, the Doppler shift in this radiation—which, like light, travels at 3×10^8 meters/sec—can easily be detected if the source or detector is moving at a millimeter per second!

These radioactive nuclei can be bound in atoms in a crystal where they vibrate back and forth in random thermal motion. Each of these nuclei forms a kind of "clock." The regular frequency emitted tells its time. If its frequency decreased, we would say it was running slow. Now, as these "clocks" vibrate back and forth in the crystal, their radiation is alternately shifted toward higher and then lower frequencies by the Doppler shift. The means by which the radiation is detected averages out these very large shifts that depend on the *direction* in which the nucleus is moving. Any effect that depends on v^2, however, does not depend on the direction of motion, but only on the magnitude of the velocity. The time dilation factor $\sqrt{1 - v^2/c^2}$ is such a term. These nuclei are really like our traveling rocket twin. They go out and come back; they accelerate, and this is really our big question. If such an accelerated clock "goes slow" by a factor $\sqrt{1 - v^2/c^2}$, i.e., by a factor that depends only on the *velocity* and has nothing to do with the *acceleration*, then our treatment of the twin paradox must be considered adequate, at least for accelerations of the size experienced by atoms in thermal vibration.

The experiment consists of measuring the frequency of light, i.e., γ-rays, emitted by these nuclei; then, warming the crystal so the nuclei move faster. The average frequency observed should decrease. Pound and Rebka [1] have observed such a decrease in frequency with increase in temperature, and the result agrees with that predicted, namely, that the accelerated clocks decrease their frequency by a factor $\sqrt{1 - v^2/c^2}$. The experiment and theory agree within the expected errors which are about ten percent.

The connection between γ-ray frequency and heartbeats may seem remote, but it is not. The γ-ray pulsations are beating out the lifetime of the radioactive nucleus, and the accelerated nuclei are "aging" more slowly than nuclei at rest, even though, on the average, they are not going anywhere. They have gone out and come back in their thermal oscillations—and they have come back younger!

Exercises

1. Repeat the calculations of Section 6-4 but in the frame of reference of the inbound pilot.

2. Suppose the traveling twin takes the trip out at a speed v, but returns at speed $v/2$. In terms of a simple time dilation calculation, how much has the traveling twin "aged" during the trip? Verify this by carrying out the Doppler effect observation of light flashes in the earth twin's reference frame. Repeat for the outbound pilot.

3. In Exercise 2, the traveling twin notices how far away the earth is just before and just after he changes rocket ships. What has happened?

[1] R. V. Pound and G. A. Rebka, *Phys. Rev. Letters* **4**, 274 (1960). A detailed discussion of this experiment as it applies to the twin paradox has been given by C. W. Sherwin, "Some Recent Experimental Tests of the Clock Paradox," *Phys. Rev.* **120**, 17 (1960).

7

The Lorentz Transformation and Notation

7-1 THE LORENTZ TRANSFORMATION

ALTHOUGH IT WOULD BE POSSIBLE to work all problems involving special relativity directly in terms of the time dilation and length contraction concepts developed in the previous chapters, it is sometimes quite inconvenient to do so. We often have to deal with events for which it is difficult to choose a reference system in which they occur at the same place and, hence, have a proper time interval to which we can apply our formulas. It is quite difficult to use our techniques in general proofs. It would be very desirable to have a set of equations, like Equation 1-1, which would tell simply and directly the relations between events as seen by two moving observers. Fortunately, we can easily obtain such a set of equations—in fact, we already have.

In Section 5-1 we considered the motion of an object from the origin $x = 0$, $y = 0$ to a general point $x = X$, $y = Y$. We considered the object to start at time $t = 0$ and arrive at $t = T$. In other words we considered two quite general events, one arbitrarily taken to be the origin in space and time, and the other taken to be any other point in space and time. As presented, the events were not completely general. For instance, we know that $\sqrt{X^2 + Y^2} < cT$ since the object cannot travel faster than light. However, even this restriction need not have been made. The reader is merely asked to review the algebra, as presented in Section 5-1, and notice that it was not really necessary to any of the arguments that the events be connected by the physical motion of an object. Consequently, these results are perfectly valid for *any* two events as seen in any two inertial frames of reference.

One of the events, the start of the trip, was taken to occur at the origin $x = y = z = 0$ at the time $t = 0$ in the frame O and also at $x' = y' = z' = t' = 0$ in the frame O'. The second event, now considered completely arbitrary, was taken to occur at $x = X$, $y = Y$, $z = Z$, and $t = T$ in O and, consequently, at a point $x' = X'$, $y' = Y'$, $z' = Z'$ at the time $t' = T'$ in O' where these are given by Equation 5-4 as

$$X' = \frac{X - vT}{\sqrt{1 - v^2/c^2}}$$
$$Y' = Y$$
$$Z' = Z$$
$$T' = \frac{T - vX/c^2}{\sqrt{1 - v^2/c^2}} \tag{7-1}$$

where the two trivial relations in y and z have been added. These equations constitute the famous Lorentz transformations of special relativity. X, Y, Z, and X', Y', Z' represent the spacial displacements, and T and T' the time interval between any two events as seen by two observers moving with respect to one another with velocity v along the x-axis.

7-2 A MATTER OF NOTATION

Up to this time, in order to leave all expressions in their most basic forms, we have avoided the introduction of a notation which is quite generally used in most writing about relativity. Every reader, at least those who have worked the exercises, will have had quite enough of writing v/c and especially $\sqrt{1 - v^2/c^2}$. We therefore introduce the following definitions:

$$\beta \equiv v/c$$
$$\gamma \equiv \frac{1}{\sqrt{1 - v^2/c^2}} \tag{7-2}$$

These have become such absolutely standard notations that physicists who deal with high energy nuclear particles use such expressions as "an electron with a beta of .85" or "a pion with a gamma of 5." The following relations, which follow quite trivially from the definitions, are often useful:

$$\gamma = \frac{1}{\sqrt{1 - \beta^2}}$$
$$\beta = \sqrt{1 - 1/\gamma^2} \tag{7-3}$$
$$\gamma^2(1 - \beta^2) = 1$$

The fact that the theory of special relativity contains a fundamental speed c makes possible two further simplifications in notation. The first of these is that the magnitude of any velocity may be written as a fraction of c. Thus, a particle which actually has a velocity uc is said to have a velocity u. This "velocity," of course, has no dimensions. The previous introduction of the symbol β was just a special case of this practice. Subsequently, in this book the general practice will be to use dimensionless velocities u, v, a, etc., when referring to the velocities of particles. The symbol β will generally be reserved for the relative velocity of two coordinate systems or observers. This artificial distinction cannot always be maintained easily and should cause no great confusion in those cases where it is not.

The fundamental speed c makes possible still another simplification in notation which makes equations in advanced relativity particularly compact: Time and distance can be measured in the same units! Such a thing is quite common in astronomy where distances are often measured in light-years, the distance light travels in a year. Physicists more commonly use a unit of length as fundamental and measure time in units of the length of time it takes light to go a unit of distance. When this is done, it is common to designate the x, y, and z coordinates by x_1, x_2, x_3, and the time coordinate of an event as $x_0 = ct$. Thus

$$x_0 = ct,\ x_1 = x,\ x_2 = y,\ x_3 = z \tag{7-4}$$

This book will use such notation sparingly. However, where time occurs in equations, it will usually be written in the combination ct, to emphasize the symmetry with which time and distance are treated in relativity.

Finally, we use the new notation to write the Lorentz transformations, Equation 7-1. We drop the capital letters, though the reader should remember that basically the equations refer to a pair of events, the first at $(0,0,0,0)$, and the second at (x,y,z,t), or (x',y',z',t'), depending on the frame of reference. The relations among the primed and unprimed coordinates of the event are:

$$\begin{aligned}x' &= \gamma(x - \beta ct)\\ y' &= y\\ z' &= z\\ ct' &= \gamma(ct - \beta x)\end{aligned} \tag{7-5}$$

Notice the complete symmetry with which x and ct are treated.

7-3 USE OF THE LORENTZ TRANSFORMATIONS

In order to understand the Lorentz transformations and what they mean, they will be applied here to various examples discussed previously.

A. *Time Dilation.* Suppose two events occur at the same place in a frame of reference O, but are separated by a time interval Δt. The first event we take to occur at the origin $x = y = z = 0$ at a time $t = 0$. The second will therefore take place at $x = y = z = 0$ at a time $t = \Delta t$. In a second frame of reference O' moving with respect to the first with a velocity βc in the x-direction, the first event can be taken to occur at $x' = y' = z' = 0$ at a time $t' = 0$. The Lorentz transformation will determine where and when the second event occurs:

$$\begin{aligned}x' &= \gamma(x - \beta ct) = \gamma(0 - \beta c\Delta t) = -\gamma\beta c\Delta t\\ y' &= y = 0\\ z' &= z = 0\\ ct' &= \gamma(ct - \beta x) = \gamma(c\Delta t - 0) = \gamma c\Delta t\end{aligned} \tag{7-6}$$

The Lorentz Transformation and Notation

The time interval, $\Delta t'$, between the events as seen in O' is, therefore, $t' - 0$ or $\gamma \Delta t$. Therefore, $\Delta t' = \gamma \Delta t$. Since Δt is a proper time interval between the events (it was in the frame of reference O that the events occurred at the same place), this is just the familiar proper-, improper-time relation. Notice that $x' = -\gamma \beta c \Delta t = -\beta c \Delta t'$. The two events occur a distance $\beta c \Delta t'$ apart in the O' frame as expected.

Although we have considered the "origin" to be the first event, it is not necessary to do so. We may do the customary thing and simply regard the origin as an arbitrary point in space and an arbitrary point in time from which all things are measured. In this second formulation we would simply pick x_1, y_1, z_1, t_1 as the coordinates of the first event, and x_2, y_2, z_2, t_2 as the coordinates of the second. But we want two events that occur at the same place, i.e., $x_1 = x_2$, $y_1 = y_2$, $z_1 = z_2$, with a time interval between them $\Delta t = t_2 - t_1$, the proper time interval between the events. Then

$$ct_1' = \gamma(ct_1 - \beta x_1)$$
$$ct_2' = \gamma(ct_2 - \beta x_2)$$
$$t_2' - t_1' = \Delta t' = \gamma(t_2 - t_1) = \gamma \Delta t \qquad (7\text{-}7)$$

Again we have $\gamma \Delta t$(proper) $= \Delta t$(improper) as required. The distance between the events in the O' system is the same as before.

B. Length Contraction. Suppose that the length of a meter stick is L measured in its own frame of reference, i.e., one end is at $x = x_A$, the other is at $x = x_B$, and $x_B - x_A = L$. This meter stick is moving past a frame of reference O' with a velocity βc. (The frame of reference O' is therefore moving with a velocity $-\beta c$ with respect to O.) An observer in O' makes a length measurement of the moving meter stick, i.e., he finds one end at x_A' and the other at x_B' *at a particular instant* $t' = 0$. The space and time coordinates of the two "measuring events" are, therefore, $x' = x_A'$, $t' = 0$ for one end, and $x' = x_B'$, $t' = 0$ for the other. The Lorentz transformation gives

$$x_A' = \gamma(x_A + \beta c t_A)$$
$$x_B' = \gamma(x_B + \beta c t_B)$$
$$t_A' = 0 = \gamma(ct_A + \beta x_A)$$
$$t_B' = 0 = \gamma(ct_B + \beta x_B)$$

where $-\beta c$ has been used for the velocity of the O' system. Solving for t_A and t_B in the last two equations, substituting in the first two, and then subtracting the first two, gives

$$x_B' - x_A' = \gamma(1 - \beta^2)(x_B - x_A) = L\sqrt{1 - \beta^2}$$

The measured distance $x_B' - x_A'$ is less than the length L by the factor $\sqrt{1 - \beta^2}$. Similarly, $t_B - t_A = -\beta(x_B - x_A)/c\sqrt{1 - \beta^2}$ which verifies the result of Section 4-4 that simultaneous events in one frame of reference are not simultaneous in another.

C. *Uniform Acceleration.* We now consider a somewhat different kind of problem, that of uniform acceleration proposed in Section 5-3. If an object is accelerated uniformly in a coordinate system O' in which it is at rest at $t' = 0$, its position is given by $x' = \frac{1}{2}a_0 t'^2$, where a_0 is its acceleration in O'. In order to find its position as a function of time in a coordinate system O, where it has a velocity vc at the time $t = 0$, we make use of the Lorentz transformation between O and O' with $\beta = v$:

$$x' = \gamma(x - \beta ct) = \frac{1}{2} a_0 t'^2$$
$$ct' = \gamma(ct - \beta x)$$

Substituting the second equation into the first gives

$$(x - \beta ct) = \frac{1}{2} a_0 \frac{\gamma}{c^2} (ct - \beta x)^2$$

This is a quadratic equation relating x and t, which, when solved for x, gives directly

$$x = \frac{c^2}{a_0 \gamma \beta^2} \left[1 + \frac{a_0 \gamma \beta t}{c} - \sqrt{\left(1 + \frac{a_0 \gamma \beta t}{c}\right)^2 - 2\frac{a_0 \gamma \beta^2}{c^2}\left(\frac{a_0 \gamma t^2}{2} + \beta ct\right)} \right]$$

By use of the relation $1 - \beta^2 = \gamma^{-2}$, this can be simplified to

$$x = \frac{c^2}{a_0 \gamma \beta^2} \left[1 + \frac{a_0 \gamma \beta t}{c} - \sqrt{1 + \frac{2a_0 \beta}{c\gamma} t} \right] \tag{7-8}$$

We have now completed our task. We know how an object which is uniformly accelerated in one frame of reference travels as a func-

The Lorentz Transformation and Notation

tion of time in another frame of reference. In one sense, Equation 7-8 may be thought of as an exact equation. *If* an object were uniformly accelerated in O', then this would be its equation of motion in O. But there is no way to make an object move with uniform acceleration, at least if its speed would thereby approach c. What we usually mean by a uniformly accelerated object is not one which has a constant acceleration in one particular inertial system, but rather one which has a constant acceleration when measured by accelerometers in its own rest frame. We may identify the accelerations measured this way with a_0 only so long as the speed of the object in O' remains small compared to c, i.e., only so long as we can apply classical ideas. This will be true only for small t' or small t. We therefore expand the radical for small t, retaining terms of order t^2, but dropping smaller ones:

$$\begin{aligned} x &= \frac{c^2}{a_0 \gamma \beta^2} \left[1 + \frac{a_0 \gamma \beta t}{c} - \left(1 + \frac{a_0 \beta}{c \gamma} t - \frac{a_0^2 \beta^2}{2 c^2 \gamma^2} t^2 + \cdots \right) \right] \\ &= \frac{c^2}{a_0 \gamma \beta^2} \left[\frac{a_0 \gamma \beta}{c} t \left(1 - \frac{1}{\gamma^2} \right) + \frac{a_0^2 \beta^2}{2 c^2 \gamma^2} t^2 + \cdots \right] \\ &= \frac{c^2}{a_0 \gamma \beta^2} \left[\frac{a_0 \gamma \beta^3}{c} t + \frac{a_0^2 \beta^2}{2 c^2 \gamma^2} t^2 + \cdots \right] \\ &= \beta c t + \frac{1}{2} \frac{a_0}{\gamma^3} t^2 + \cdots \end{aligned}$$

At $t' = 0$, and hence $t = 0$, the object had zero velocity in the primed system and, therefore, has a velocity βc in the unprimed system. It had an acceleration a_0 in the primed system and an acceleration $a_0 \gamma^{-3}$ in the unprimed system. This agrees exactly with Equation 5-11.

Exercises

1. Solve the Lorentz transformation equations, Equation 7-1, for the unprimed in terms of the primed coordinates. How do you know you are right?

2. If $U_x = dx/dt$, $U_y = dy/dt$, and $U_x' = dx'/dt'$, $U_y' = dy'/dt'$, derive the velocity addition formulas by direct differentiation of the Lorentz transformation equations.

3. Do Exercise 3, in Chapter 4, by means of the Lorentz transformations, taking the flash F to be the origin $x = x' = t = t' = 0$.

4. A piece of glass of thickness s and index of refraction n is moving through the laboratory at a speed v. A flash of light passes through the glass in the direction in which the glass is moving. In the laboratory frame of reference, how long did it take the light to pass through the glass? How far apart, in the laboratory, were the points where the light entered and left the glass? (The index of refraction n means that the speed of light in the glass is c/n when measured in the frame of reference of the glass.)

5. Two rocket ships have identical lengths L as measured in their own frames of reference. In an inertial frame of reference O these rocket ships pass in opposite directions, one at a velocity $2v$, the other at $-v$. How far apart in space and in time are the events when the two prows pass and the two sterns pass? Calculate these in all three reference frames: O, and both rockets. How do the results in the two rocket frames compare? How *should* they compare?

6. A horizontal meter stick of length L drops vertically to the floor with a speed uc. (Don't worry about accelerations.) The collisions of its ends with the floor are simultaneous events. To an observer running along the floor at speed βc the meter stick appears to drop in a non-vertical direction.

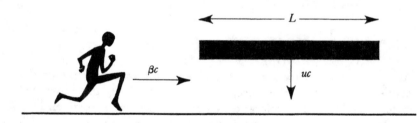

a. Calculate the tangent of the angle to the vertical at which the meter stick appears to drop, i.e., the direction of its velocity as seen by the running observer.

b. Furthermore, to the moving observer, the collisions of the ends with the floor are not simultaneous. How far apart do they occur in time?

c. In view of (*b*), with what angle to the floor does the meter stick appear to be tilted?

d. For what speed uc does the meter stick remain perpendicular to its direction of motion for all speeds βc of the observer?

7. Use the Lorentz transformations directly to show that if two events

The Lorentz Transformation and Notation

are "connected by a light flash" in one frame of reference, they are also in any other. That is, if in one frame of reference

$$c(t_B - t_A) = \sqrt{(x_B - x_A)^2 + (y_B - y_A)^2}$$

Then, in another

$$c(t_B' - t_A') = \sqrt{(x_B' - x_A')^2 + (y_B' - y_A')^2}$$

8. *a.* If one event occurs at (x_A, y_A) at the time t_A in a particular frame of reference, and another at (x_B, y_B, t_B), show that the proper time Δt between these events is given by

$$c^2 \Delta t^2 = c^2(t_B - t_A)^2 - (x_B - x_A)^2 - (y_B - y_A)^2$$

b. In a second frame of reference where the positions and times of these events are x_A', y_A', t_A', and x_B', y_B', t_B' the proper time interval is, of course, given by the same formula with primes. Use the Lorentz transformation to show that the proper time interval between these events is the *same* as calculated in both frames of reference. Should it be? That is, is the proper time interval between events something that is a unique property of the events, or does it depend on the frame of reference from which they are viewed?

9. If no signal can travel faster than the speed of light, then, for one event A to be the cause of another event B, or B the cause of A, the time interval between B and A must be greater than the time it takes light to go from A to B. Thus

$$|t_B - t_A| > |x_B - x_A|/c \quad \text{or} \quad (t_B - t_A)^2 > (x_B - x_A)^2/c^2 \tag{1}$$

where the two events have both been chosen to lie along the x-axis.

a. Show that if Equation *1* is true in one frame of reference, it is true in all frames moving uniformly with respect to it.

b. Show that if (*1*) is true, there is at least one reference frame where the interval $t_B - t_A$ is proper, and there is no reference frame where events A and B are simultaneous.

c. If on the other hand,

$$|t_B - t_A| < |x_B - x_A|/c \tag{2}$$

show that there is a reference frame where the events are simultaneous, but none where the time interval is proper.

d. The interval between events related like (*1*) is called *timelike* since there is one reference frame where the interval is a "pure" time. Events like (*2*) are said to be *spacelike*. Discuss the fact that the property of being timelike or spacelike is invariant, i.e., does not depend on which reference frame is used to describe them. Particularly, relate your discussion to causality, i.e., whether or not one event can cause a second.

10. Why is Equation 7-8 different from the equation at the end of Section 5-3?

11. Suppose an object has an acceleration a_0 as measured in its own reference frame but perpendicular to its direction of motion through a system O with speed βc. What is its acceleration as measured in O?

8

Proper- or Four-Velocity

8-1 TWO KINDS OF VELOCITY

WHEN A MOTORIST WISHES to determine his average speed between two cities, which are, for example, 80 miles apart, he divides the distance traveled by the time it took. If it took him 2 hr, his speed was 40 mi/hr. No one bothers to ask him how he measured time. He may, on the one hand, have noticed that he left the first city at 10 A.M. and arrived at the second at 12 noon by looking at clocks in the cities as he passed by. On the other hand, he may have timed the trip by means of his dashboard clock. Classically, these are equally valid speed measurements. We see, however, that the dashboard clock measures the proper time interval between the start and end of the trip, whereas clocks stationary on the ground give an improper time interval. When speeds are large these do not agree. Which determination of speed or velocity is valid for high speeds?

Actually both kinds of velocity are useful in relativistic calculations. Perhaps the most natural choice for the velocity of an object in a particular frame of reference is the distance of the trip in that reference frame, divided by the time interval as measured by clocks at rest in that reference frame, i.e., the improper time. This is the quantity that we have, up to now, called the velocity, which is the generally accepted practice.

However, the other possibility is sometimes useful, and in many respects turns out to be a more natural choice for the relativistic generalization of velocity. Remember that *both* choices are acceptable classical definitions; at low speeds both will give the same result. The choice of which to call *the* velocity of a high speed object is dictated by convenience and tradition only.

To obtain the relation between these two kinds of velocity—the distance divided by the proper time is called the *proper velocity*—consider the object in Figure 8-1 that moves from the origin to the point (X,Y) in the time T as measured by clocks at rest in the O coordinate system. The (ordinary) velocity components are given by

$$U_x = X/cT \qquad U_y = Y/cT \qquad U^2 = U_x^2 + U_y^2 \tag{8-1}$$

Notice that the velocities of the object are given in dimensionless units by dividing by c. The proper time interval for the trip is measured by a clock on the object and is given by

$$\tau = T\sqrt{1 - U^2}$$

The *proper velocity* of the object is defined as the distance traveled divided by the proper time. It is usual here, also, to divide by c so

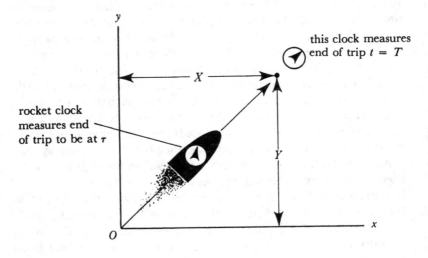

Figure 8-1

Proper- or Four-Velocity

that the proper velocity is expressed as a ratio to the speed of light. As often as practical we shall use the symbol η for the proper velocity. We define

$$\eta_x \equiv \frac{X}{cT\sqrt{1-U^2}} = \frac{U_x}{\sqrt{1-U^2}}$$
$$\eta_y \equiv \frac{U_y}{\sqrt{1-U^2}}$$
(8-2)

These equations give the desired relation between the proper velocity and the velocity of an object.

We shall see in a later chapter that, unlike the velocity, the proper velocity forms part of a four-vector and is sometimes called the *four-velocity*.

Not only does the proper velocity reduce to the classical velocity at low speeds, but at the speed of light it becomes infinite. That the ultimate speed becomes infinite satisfies our intuition better than that it should reach a finite limit. However, much the most important property of the proper velocity is how it changes for different observers.

8-2 ADDITION OF VELOCITIES FORMULA FOR PROPER VELOCITY

Suppose one observer O sees an object with velocity components U_x, U_y, (Figure 8-2a) and proper velocity components η_x, η_y given by Equation 8-2 (Figure 8-2b). Another observer O' is moving to the right with respect to O with a speed βc. He therefore measures related velocity components U_x' and U_y' of the object (Figure 8-2c), which are given by Equation 5-5. Our present question is: What are the proper velocity components η_x' and η_y' that O' observes? There are two quite distinct processes by which the question can be answered, and each is informative in its way.

A. *Direct Calculation.* This first method of calculation is not at all clever. In fact, the algebra is exceedingly messy. It is given here explicitly to show that such direct calculations with relativistic formulas are possible, and that the results, which are very involved at

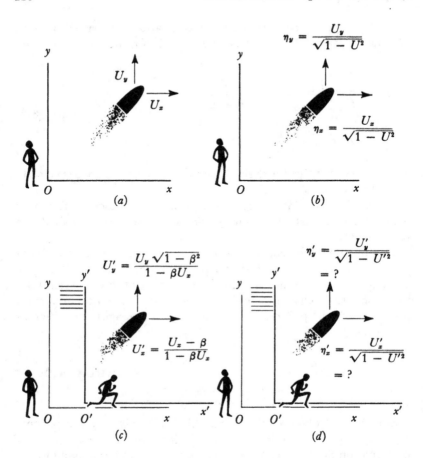

Figure 8-2. (a) *Suppose an observer O sees an object moving with velocity components U_x and U_y.* (b) *The observer O might wish to talk about proper velocity components η_x and η_y, instead.* (c) *An observer O' moving at speed βc with respect to O, would see the object moving with velocity components U_x' and U_y' given by the velocity addition formulas.* (d) *On the other hand, the observer O' might wish to use η_x' and η_y' instead. He knows how to get η_x' and η_y' in terms of U_x' and U_y'. The question is: How are η_x' and η_y' related to proper velocity components seen by O?*

Proper- or Four-Velocity

intermediate stages, when finally resolved are quite simple. This is quite characteristic of relativistic calculations.

From Equation 8-2 the proper velocity in the O' frame of reference must be given by

$$\eta_x' = \frac{U_x'}{\sqrt{1 - U'^2}} \qquad \eta_y' = \frac{U_y'}{\sqrt{1 - U'^2}}$$

On the other hand, from Equation 5-5, but using velocities expressed as fractions of c

$$U_x' = \frac{U_x - \beta}{1 - U_x\beta} \qquad U_y' = \frac{U_y \sqrt{1 - \beta^2}}{1 - U_x\beta}$$

Substitution of the second set of equations into the first will give the answer. We prefer to do the calculation in steps, first calculating

$$\frac{1}{\sqrt{1 - U'^2}} = \frac{1}{\sqrt{1 - \left[\left(\frac{U_x - \beta}{1 - U_x\beta}\right)^2 + \left(\frac{U_y \sqrt{1 - \beta^2}}{1 - U_x\beta}\right)^2\right]}}$$

$$= \frac{(1 - U_x\beta)}{\sqrt{(1 - U_x\beta)^2 - (U_x - \beta)^2 - U_y^2(1 - \beta^2)}}$$

$$= \frac{(1 - U_x\beta)}{\sqrt{1 - 2U_x\beta + U_x^2\beta^2 - U_x^2 + 2U_x\beta - \beta^2 - U_y^2 + U_y^2\beta^2}}$$

$$= \frac{(1 - U_x\beta)}{\sqrt{1 - (U_x^2 + U_y^2) - \beta^2 + (U_x^2 + U_y^2)\beta^2}}$$

$$= \frac{(1 - U_x\beta)}{\sqrt{1 - U^2 - \beta^2 + U^2\beta^2}}$$

$$= \frac{(1 - U_x\beta)}{\sqrt{(1 - U^2) - \beta^2(1 - U^2)}}$$

$$= \frac{(1 - U_x\beta)}{\sqrt{(1 - \beta^2)(1 - U^2)}}$$

$$= \frac{1 - U_x\beta}{\sqrt{1 - \beta^2}\sqrt{1 - U^2}} \qquad (8\text{-}3)$$

It is now easy to find η_x' and η_y'.

$$\eta_x' \equiv \frac{U_x'}{\sqrt{1-U'^2}} = \frac{\dfrac{U_x - \beta}{1 - U_x\beta}(1 - U_x\beta)}{\sqrt{1-\beta^2}\sqrt{1-U^2}}$$

$$= \frac{1}{\sqrt{1-\beta^2}}\left[\frac{U_x}{\sqrt{1-U^2}} - \frac{\beta}{\sqrt{1-U^2}}\right] \quad (8\text{-}4)$$

$$\eta_y' \equiv \frac{U_y'}{\sqrt{1-U'^2}} = \frac{\dfrac{U_y\sqrt{1-\beta^2}}{1-U_x\beta}(1-U_x\beta)}{\sqrt{1-\beta^2}\sqrt{1-U^2}}$$

$$= \frac{U_y}{\sqrt{1-U^2}} \quad (8\text{-}5)$$

For convenience (though this will be seen to have considerable significance later) we define

$$\eta_0 = \frac{1}{\sqrt{1-U^2}}$$

We now write Equations 8-3, 8-4, and 8-5 in terms of η's alone:

$$\eta_0' = \frac{\eta_0 - \eta_x\beta}{\sqrt{1-\beta^2}} = \gamma(\eta_0 - \beta\eta_x)$$

$$\eta_x' = \frac{\eta_x - \eta_0\beta}{\sqrt{1-\beta^2}} = \gamma(\eta_x - \beta\eta_0) \quad (8\text{-}6)$$

$$\eta_y' = \eta_y$$

The remarkable thing about Equation 8-6 is that it looks exactly like the Lorentz transformations for x, y, and t provided that ct is equivalent to η_0, x to η_x, and y to η_y.

The reader, no doubt, found the previous work quite complicated algebraically. It should be pointed out that, although complicated, it was extremely straightforward. If we are given some algebraic quantity—in our case $\eta_x' = U_x'/\sqrt{1-U'^2}$—and we want to find

Proper- or Four-Velocity

out what η_x' is in terms of U_x and U_y, we simply substitute for U_x' and U_y' in terms of U_x and U_y by means of the previously calculated velocity addition formulas, and grind the algebraic crank. Finally, we prefer to know η_x' in terms of η_x, etc., rather than the regular velocities U_x. That the final result is so simple is due to the very special algebraic form of η. We will see in a later chapter that quantities which transform from one frame of reference to another in this simple way have a special significance in relativity and are called *four-vectors*.

B. *Calculation by Definition.* Part A of this section was a tedious algebraic manipulation done in order to relate the proper velocity of an object in one frame of reference to that in another frame of reference. It must have seemed almost a complete algebraic accident to most readers that the final form turned out to be closely related to the Lorentz transformation. In this part, the calculation is repeated in a way that will show more directly why this is so.

Figure 5-2 shows an object moving from the origin to a final position (X,Y) as measured in O and (X',Y') as measured in O'. In the analysis of Chapter 5 we calculated the two velocities by dividing the distance traveled by the time as measured with clocks at rest in the O and O' systems respectively, i.e., $U_x = X/cT$ and $U_x' = X'/cT'$. We now wish to calculate the *proper velocities* η_x and η_x', which are the distances traveled in O and O' respectively, divided by the *proper time* interval for the trip. Let us call this proper time for the trip τ. We then have

$$\eta_x = \frac{X}{c\tau} \qquad \eta_y = \frac{Y}{c\tau}$$

$$\eta_x' = \frac{X'}{c\tau} \qquad \eta_y' = \frac{Y'}{c\tau} \tag{8-7}$$

Notice that the proper time for the trip is measured neither with clocks in the O system nor with clocks in the O' system. It is a time interval measured with clocks on the object. An observer in O and an observer in O' get the same answers for the time τ, since they could both obtain this answer, for instance, by photographing the clocks on the object at the start and finish with cameras on the object. They would both compute τ by looking at the same set of pictures and, clearly, would get the same answers. Alternatively the pilot of the rocket ship could radio back to O and O' the time τ for the trip. Obviously, relativistic considerations are not going to garble the language of his message so that he appears to say "3 hours" when heard by O, and "4 hours" when heard by O'! The *proper time* for the trip (or between any two events for that matter) is a definite quantity and is the same for observers in *any* reference system. (Compare also the results of Exercise 8.) Such a quantity is called a *four-scalar*, but we need not be concerned with this vocabulary just yet. The whole point is that we *don't* use τ in O and τ' in O'; the *proper time* is the same for both frames of reference.

Since τ is the same for both frames of reference, it is no longer surprising that η_x and η_y should transform into η_x' and η_y' like X and Y into X' and Y' since they are just a constant times X and Y! In order to carry through the entire process of getting η_x' in terms of η_x we must apply the Lorentz transformations to X' and Y'. The y component is trivial

$$\eta_y' = \frac{Y'}{c\tau} = \frac{Y}{c\tau} = \eta_y$$

The x-component gives

$$\eta_x' = \frac{X'}{c\tau} = \frac{\gamma(X - \beta c T)}{c\tau} = \gamma\left(\frac{X}{c\tau} - \beta\frac{cT}{c\tau}\right) = \gamma\left(\eta_x - \beta\frac{T}{\tau}\right)$$

If we now define a quantity $\eta_0 = cT/c\tau$ in complete accordance with $\eta_x = X/c\tau$, etc., we have

$$\begin{aligned}\eta_x' &= \gamma(\eta_x - \beta\eta_0) \\ \eta_y' &= \eta_y\end{aligned} \tag{8-8}$$

These are, of course, the results given by Equation 8-6, except that no transformation is given for η_0' there. But with our present understanding of η_0 and the Lorentz transformation of the time, we obtain this immediately as

$$\eta_0' = \frac{cT'}{c\tau} = \frac{\gamma(cT - \beta X)}{c\tau} = \gamma(\eta_0 - \beta\eta_x)$$

8-3 AN EXAMPLE

The preceding work has been rather formal, and it is suggested that the reader fill in the blanks in the following exercise. (Answers are at the end of the chapter.)

A muon traveling along the x-axis in the laboratory decays into an electron and two neutrinos. In the frame of reference of the muon, the electron starts from the origin $x = 0, y = 0$ at time $t = 0$ and arrives at a point $x = 36$ meters, and $y = 27$ meters at a time $t = 0.25$ microseconds. What are its velocity components in the frame of reference of the muon? $u_x = $ _____, $u_y = $ _____. (a) What is its speed? $u = $ _____. (b) According to clocks on the electron (!), how long did the trip take? $t_e = $ _____. (c) Calculate the proper velocity components of the electron in the frame of reference of the muon from its velocity components. $\eta_x = $ _____, $\eta_y = $ _____. (d) Calculate the proper velocity components directly as the distance traveled divided by the proper time for the trip. $\eta_x = $ _____, $\eta_y = $ _____. (e) Calculate the "time component" of the proper velocity, $\eta_0 = $ _____. (f)

Now when it decayed, the muon was traveling at a speed $4c/5$ toward positive x. The laboratory is therefore traveling toward negative x with the same speed with respect to the muon. By use of the ordinary relativistic velocity addition equations, find the velocity components of the electron with respect to the laboratory. $u_x' = $ _____, $u_y' = $ _____. (g) From these find its proper velocity components in the laboratory. $\eta_x' = $ _____, $\eta_y' = $ _____. (h) Now, instead of using this last method, use the Lorentz transformations to find the space and time position of the beginning and end of the trip. The start is at $x' = y' = t' = 0$. The end is at $x' = $ _____, $y' = $ _____, $t' = $ _____. (i) In this frame of reference calculate the proper velocity directly. $\eta_x' = $ _____, $\eta_y' = $ _____, $\eta_0' = $ _____. (j) Finally, find the proper velocity components by performing a Lorentz transformation on the proper velocity components in the muon rest frame, (e) and (f) above. $\eta_x' = $ _____, $\eta_y' = $ _____, $\eta_0' = $ _____. (k)

Exercises

1. An astronaut sets off for α-Centauri, four light-years distant, at a proper velocity of 20 with respect to the earth. When does he get there according to his own clocks?

2. If an object, A, has a velocity $c/\sqrt{2}$ at 45° to the x-axis, what are its proper velocity components? If a second object, B, had the same speed at 135° to the x-axis, what are its proper velocity components?

3. What are the proper velocity components of the two objects of Exercise 2 in a coordinate system moving along the positive x-axis with a speed $c/2$ with respect to the one of Exercise 2?

4. *a.* Suppose you call the proper velocity components found in Exercise 2 $\eta_{xA}, \eta_{yA}, \eta_{0A}$ and $\eta_{xB}, \eta_{yB}, \eta_{0B}$ and those of Exercise 3 $\eta_{xA}', \eta_{yA}', \eta_{0A}'$ and $\eta_{xB}', \eta_{yB}', \eta_{0B}'$. Show by simply substituting the numbers that $\eta_{0A}\eta_{0B} - \eta_{xA}\eta_{xB} - \eta_{yA}\eta_{yB} = \eta_{0A}'\eta_{0B}' - \eta_{xA}'\eta_{xB}' - \eta_{yA}'\eta_{yB}'$.

b. Prove that for any two objects and any two inertial frames of reference $\eta_{0A}\eta_{0B} - \eta_{xA}\eta_{xB} - \eta_{yA}\eta_{yB} = \eta_{0A}'\eta_{0B}' - \eta_{xA}'\eta_{xB}' - \eta_{yA}'\eta_{yB}'$.

5. Show that if $\eta_{xC} = \eta_{xA} + \eta_{xB}$, $\eta_{yC} = \eta_{yA} + \eta_{yB}$ and $\eta_{0C} = \eta_{0A} + \eta_{0B}$, that $\eta_{xC}' = \eta_{xA}' + \eta_{xB}'$, etc. Is this true of ordinary velocities, i.e., if $U_{xC} = U_{xA} + U_{xB}$, etc., is $U_{xC}' = U_{xA}' + U_{xB}'$?

Answers to Completion Sentences on Page 121

a. $u_x = 0.48 \quad u_y = 0.36 \quad$ (given as fractions of c)

b. $u = 0.60$

c. $t_e = $ (proper time for the trip) $= 0.20 \times 10^{-6}$ sec

d. $\eta_x = \dfrac{u_x}{\sqrt{1-u^2}} = \dfrac{.48}{\sqrt{1-(.6)^2}} = 0.6 \quad \eta_y = 0.45$

e. $\eta_x = \dfrac{36 \text{ meters}}{.2 \times 10^{-6} \times 3 \times 10^8} = .6 \quad \eta_y = 0.45$

f. $\eta_0 = \dfrac{ct}{ct_e} = \dfrac{0.25}{0.20} = 1.25$

g. $u_x' = \dfrac{u_x - \beta}{1 - \beta u_x} = \dfrac{.48 + .8}{1 + .8 \times .48} = \dfrac{160}{173} = .922 \quad u_y' = \dfrac{27}{173} = .156$

h. $\eta_x' = \dfrac{u_x'}{\sqrt{1 - u_x'^2 - u_y'^2}} = \dfrac{\left(\dfrac{160}{173}\right)}{\sqrt{1 - \left(\dfrac{160}{173}\right)^2 - \left(\dfrac{27}{173}\right)^2}}$

$= \dfrac{8}{3} \quad \eta_y' = \dfrac{9}{20} = .45$

Proper- or Four-Velocity

Note that $\eta_y' = \eta_y$ as it should.

i. $x' = \gamma(x - \beta ct) = \dfrac{1}{\sqrt{1 - \left(\dfrac{4}{5}\right)^2}} \left(36 + \dfrac{4}{5} \times 3 \times 10^8 \times .25 \times 10^{-6}\right)$

$= 160$ meters $\quad y' = y = 27$ meters

$ct' = \gamma(ct - \beta x) = \dfrac{1}{\sqrt{1 - \left(\dfrac{4}{5}\right)^2}} \left(3 \times 10^8 \times .25 \times 10^{-6} + \dfrac{4}{5} \times 36\right)$

$= 173$ meters

j. $\eta_x' = \dfrac{x'}{c\tau} = \dfrac{160}{3 \times 10^8 \times .20 \times 10^{-6}} = \dfrac{8}{3}$

$\eta_y' = \dfrac{y'}{c\tau} = \dfrac{9}{20} \qquad \eta_0' = \dfrac{ct'}{c\tau} = \dfrac{173}{60}$

k. $\eta_x' = \gamma(\eta_x - \beta \eta_0) = \dfrac{5}{3}(.6 + .8 \times 1.25) = \dfrac{8}{3}$

$\eta_y' = \eta_y = \dfrac{9}{20} \qquad \eta_0' = \gamma(\eta_0 - \beta \eta_x) = \dfrac{173}{60}$

9

Momentum and Energy

9-1 NON-RELATIVISTIC CONSERVATION OF MOMENTUM

BEFORE TURNING TO THE CONSIDERATION of momentum and energy in relativistic mechanics, we will find it instructive to review a few of the ideas about non-relativistic conservation of momentum which we first met in Chapter 1.

A very general collision between two particles is shown in Figure 9-1a. Here an object of mass m_a and velocity $\mathbf{a}c$ collides with an object of mass m_b and velocity $\mathbf{b}c$ to produce two objects of mass m_d and m_e with velocities $\mathbf{d}c$ and $\mathbf{e}c$. (A vector will be denoted by bold face type, its magnitude by the same symbol in italics, and its components by the italic symbol with subscripts.) No great significance need be read into the use of different incoming and outgoing masses. The collision might simply be one where a shovelful of sand was dumped from one cart into another. The velocities have been written in terms of c even in this non-relativistic case, because this same figure and the same velocities will be used relativistically later on where this will simplify the formulas. In Figure 9-1b the *same* collision is viewed by an observer moving to the right with a velocity βc. In this second frame of reference the colliding objects have velocities $\mathbf{a}'c$, $\mathbf{b}'c$, $\mathbf{d}'c$, and $\mathbf{e}'c$.

Momentum and Energy

If the observer of Figure 9-1b is in an inertial reference frame he finds momentum conserved, and may write for x- and y-components (the z-component is ignored for simplicity)

$$m_a a_x' + m_b b_x' = m_d d_x' + m_e e_x'$$
$$m_a a_y' + m_b b_y' = m_d d_y' + m_e e_y' \qquad (9\text{-}1)$$

If the non-relativistic transformation of velocities is assumed, $a_y' = a_y$, etc., the y-component of Equation 9-1 becomes

$$m_a a_y + m_b b_y = m_d d_y + m_e e_y \qquad (9\text{-}2)$$

This equation, of course, simply expresses momentum conservation in the frame of reference of Figure 9-1a. If momentum is conserved in one inertial frame of reference, it is conserved in all.

Only the y-component was considered in the last paragraph. Still, the result is perfectly general. The direction chosen for the x- or y-axes is completely arbitrary. If one component of momentum is conserved, then all must be.

The velocity transformation in the x direction is slightly more complicated: $a_x' = a_x - \beta$, etc. If this is substituted in the x-component of Equation 9-1 it becomes

$$m_a(a_x - \beta) + m_b(b_x - \beta) = m_d(d_x - \beta) + m_e(e_x - \beta)$$

or

$$m_a a_x + m_b b_x - m_d d_x - m_e e_x + \beta(m_a + m_b - m_d - m_e) = 0 \qquad (9\text{-}3)$$

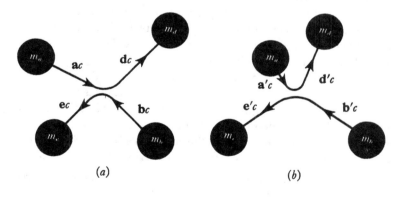

Figure 9-1

Since β is a completely arbitrarily chosen velocity, Equation 9-3 can be true only if

$$m_a a_x + m_b b_x = m_d d_x + m_e e_x \quad \text{and} \quad m_a + m_b = m_d + m_e \quad (9\text{-}4)$$

In other words, if momentum conservation is to be a valid non-relativistic law, i.e., valid in all inertial systems, then not only must momentum be conserved in a collision, but also the sum of the masses before and after the collision must be the same, i.e., mass must be conserved. Although mass conservation was probably tacitly assumed even before the law of conservation of momentum was recognized, this argument shows that it could have been deduced from the momentum conservation law and the principle that the law must hold in all inertial frames, i.e., the principle of relativity.

9-2 HOW DO WE CHOOSE A RELATIVISTIC EXPRESSION FOR MOMENTUM?*

The velocity addition formulas, which are good at high speeds, are derived from the Lorentz transformation and not the non-relativistic transformation of Equation 1-1. It would be surprising, therefore, if the classical momentum, mass times velocity, went over unmodified into relativistic mechanics. If the momentum conservation equations (Equation 9-1) are written down in terms of the unprimed velocities by means of the velocity addition formulas (Equation 5-5), no such simple relations as Equations 9-2 and 9-3 ensue. It becomes evident very quickly that conservation of "mass times velocity" cannot be a physical law valid for all observers.

The classical ideas of momentum conservation fail. Where do we go from here? It may be that there is nothing like classical conservation of momentum for objects moving near the speed of light; it may be that the classical scheme of mechanics must be completely revised. On the other hand, it may simply be that there is a different expression for momentum than the one we have tried. But if this latter is

---

* The argument given in this section is an adaption of that given by W. Pauli, *Theory of Relativity* (Pergamon Press, New York, 1958) p. 118, which was taken from G. M. Lewis and C. Tolman, *Phil. Mag.*, 18, 510 (1909).

Momentum and Energy

correct, how can the new expression be derived? The answer is that it cannot! It must be guessed! This does not mean that one expression after another must be tried at random; several things can guide the choice. Foremost of these is that any new expression must become the classical value for small velocities, for there is a wealth of information that tells us "mass times velocity" is conserved for slowly moving objects. Secondly, such an expression must give a law of conservation of momentum that is valid in all inertial frames of reference. It should be remembered that Einstein also had a knowledge of electricity and magnetism that is beyond the scope of this book. Certain ideas from electromagnetic fields are very suggestive of the proper choice.

We choose to consider a glancing collision between a rapidly moving object and a slowly moving one of equal mass. This collision is shown in Figure 9-2a. An object moves in from the top with a high velocity uc and bounces symmetrically off another object of equal mass that comes in from the bottom, vertically with velocity vc. Since the collision is symmetrical, the second object bounces directly down again with the velocity $-vc$. A very glancing collision is considered because although uc can be large, if θ is very small, vc can be so small that the momentum of the lower object can be treated classically. It is in this way that we make a connection between relativistic and classical momenta.

Suppose that the rapidly moving object has a momentum **p**. Since we still presume that momentum is a vector, its vertical component is $p \sin \theta$. If momentum is to be conserved in the collision

$$2p \sin \theta = 2mvc \quad \text{or} \quad p \sin \theta = mcv \tag{9-5}$$

The horizontal component of momentum is clearly conserved because of the symmetry.

Figure 9-2b shows the same collision from the point of view of an observer moving to the right with a speed $uc \cos \theta$, i.e., so as to just keep pace with the upper mass in its horizontal motion. To this observer, the upper mass comes in with velocity $\mathbf{u}'c$. The velocity addition formula, Equation 5-5 gives

$$u'c = \frac{uc \sin \theta \sqrt{1 - u^2 \cos^2 \theta}}{1 - (u \cos \theta)(u \cos \theta)} = \frac{uc \sin \theta}{\sqrt{1 - u^2 \cos^2 \theta}} \tag{9-6}$$

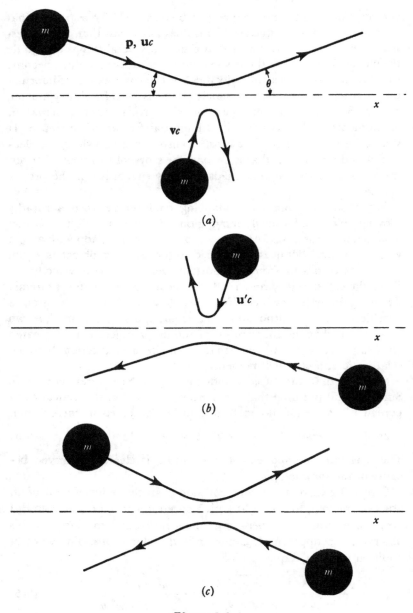

Figure 9-2

Momentum and Energy

However, the symmetry of the collision tells us that $u'c = vc$, i.e., that Figure 9-2b is just the same as Figure 9-2a turned upside down. If it is not clear to the reader that this collision must be so symmetrical, consider that there must be some frame of reference where the collision appears as in Figure 9-2c, a completely symmetrical case. Substitution of Equation 9-6 into 9-5 gives

$$p \sin \theta = mcv = mcu' = mc \frac{u \sin \theta}{\sqrt{1 - u^2 \cos^2 \theta}}$$

which gives

$$p = \frac{mcu}{\sqrt{1 - u^2 \cos^2 \theta}} \tag{9-7}$$

In arriving at Equation 9-7 it was necessary to assume that θ was small in order that the collision should leave one object with a small enough speed to be treated non-relativistically. The approximation should become better as θ approaches zero. When θ is zero, Equation 9-7 becomes

$$p = \frac{mcu}{\sqrt{1 - u^2}} \quad \text{or in vector form} \quad \mathbf{p} = \frac{mc\mathbf{u}}{\sqrt{1 - u^2}} \tag{9-8}$$

Equation 9-8 turns out to be the correct relativistic expression for momentum. In the next sections it will be investigated further. For now note that it has at least one property which it should. At low velocities it reduces to the classical "mass times velocity."

Another interesting property of Equation 9-8 is that the relativistic momentum is seen to be given by the *"mass times proper velocity."* The proper velocity of an object, it should be remembered, is the distance traveled by the object divided by the time for the trip as measured by a clock carried on the object. This is a definition of velocity that is, classically, equally as valid as that of the ordinary velocity. In fact, having failed with "mass times ordinary velocity" as a formula for momentum, "mass times proper velocity" is probably the next logical choice for trial even without the previous discussion of bouncing masses!

In arriving at the momentum expression, Equation 9-8, one of the objects was treated classically. If Equation 9-8 is completely correct it can be used for both particles without approximation. The reader is urged to try this as follows:

In Figure 9-2a a particle of speed uc comes in at an angle θ (which is no longer required to be small) and collides with a particle moving vertically at a speed vc. We suppose, as before, that the particle of speed uc has a momentum $\mathbf{p}$. We again require momentum conservation in the y-direction. Now, however, the relativistic expression Equation 9-8 is used for the momentum of the particle of speed vc. We get $p \sin \theta =$ _____. (a) We look at the collision in a frame of reference traveling to the right at a speed _____. (b) In this frame of reference, the upper particle has zero x-component of velocity, and its y-component is $u'c =$ _____. (c) Since, by symmetry, this must be the same as the speed vc, we set $vc = u'c$ and substitute into expression (a). This gives $p =$ _____, (d), which is exactly the desired expression for momentum. Our expression, when used consistently, then, is in no way an approximation. (Answers are at the end of the chapter.)

9-3 RELATIVISTIC MOMENTUM CONSERVATION

The argument of the last section that Equation 9-8 represents the correct relativistic expression for momentum can hardly be considered rigorous in view of the particularly symmetric collision considered there. Nevertheless, it can serve as motivation for testing whether that expression can be a valid relativistic expression for momentum. We return to the consideration of the very general collision of Figure 9-1 where two objects with masses m_a and m_b collide with velocities $\mathbf{a}c$ and $\mathbf{b}c$ to produce two objects of masses m_d and m_e leaving with velocities $\mathbf{d}c$ and $\mathbf{e}c$.

We proceed exactly as in Section 9-1, asking under what conditions the observer of Figure 9-1b finds the y-component of momentum conserved. He would express conservation of momentum as

$$p_{ay}' + p_{by}' = p_{dy}' + p_{ey}' \tag{9-9}$$

Momentum and Energy

which, using the expression (Equation 9-8) for momentum, means

$$\frac{m_a c a_y'}{\sqrt{1-a'^2}} + \frac{m_b c b_y'}{\sqrt{1-b'^2}} = \frac{m_d c d_y'}{\sqrt{1-d'^2}} + \frac{m_e c e_y'}{\sqrt{1-e'^2}} \qquad (9\text{-}10)\ ^1$$

In order to express Equation 9-10 in terms of the unprimed velocities of Figure 9-1a, the primed velocities could be written in terms of the unprimed by means of the velocity addition formulas; but, this would merely be repeating the laborious work of Chapter 8 on proper velocities. It is more convenient to make use of Equation 8-1 and write Equation 9-10 directly in terms of the proper velocities:

$$m_a \eta_{ay}' + m_b \eta_{by}' = m_d \eta_{dy}' + m_e \eta_{ey}' \qquad (9\text{-}11)$$

However, the y-component of proper velocity does not change when observed by an observer moving along the x-axis (see Equation 8-6) and the conservation of momentum, Equation 9-11, becomes

$$m_a \eta_{ay} + m_b \eta_{by} = m_d \eta_{dy} + m_e \eta_{ey}$$

This simply expresses conservation of momentum in the frame of reference of Figure 9-1a. It shows that if the y-component of momentum is conserved in one frame of reference, it is conserved in all frames of reference moving uniformly with respect to the first.

[1] Notice that to take a component of momentum we do it in the following way: The magnitude of the momentum is

$$p = \frac{mcu}{\sqrt{1-u^2}}$$

The components are therefore

$$p_x = p \cos \theta = \frac{mcu \cos \theta}{\sqrt{1-u^2}} = \frac{mcu_x}{\sqrt{1-u^2}}$$

$$p_y = p \sin \theta = \frac{mcu \sin \theta}{\sqrt{1-u^2}} = \frac{mcu_y}{\sqrt{1-u^2}}$$

The x-component of the momentum is *not*

$$p_x = \frac{mcu_x}{\sqrt{1-u_x^2}} \longleftarrow \text{(NOT } x\text{)}$$

So far only the y-component of momentum has been considered. However, what is called the y-component is completely arbitrary. If conservation of one component of momentum can be shown to be a physical law, then it follows that all components are conserved. It must be remembered that what has been shown is that conservation of momentum as expressed by Equation 9-8 does not violate the first postulate of relativity. That is, if one observer finds it to be true, then all observers will do so. Whether or not conservation of this quantity is actually observed in nature must be left to experiment to decide.

9-4 A NEW CONSERVATION LAW: ENERGY

Having decided that conservation of momentum, as expressed by Equation 9-8, does not violate the relativity postulate, and deferring the discussion of experimental verification, we proceed to investigate the consequences of this law by considering how two observers would describe the conservation of momentum along the direction of their relative motion, the x-direction.

The observer of Figure 9-1b would write down conservation of momentum in the x-direction as

$$p_{ax}' + p_{bx}' = p_{dx}' + p_{ex}'$$

or in terms of the proper velocities

$$m_a \eta_{ax}' + m_b \eta_{bx}' = m_d \eta_{dx}' + m_e \eta_{ex}' \tag{9-12}$$

In order to see what this implies for the observer of Figure 9-1a, this must be written in terms of velocities as measured by him. The transformation of the x-component of proper velocity is given by Equation 8-6:

$$\eta_{ax}' = \frac{a_x'}{\sqrt{1-a'^2}} = \frac{1}{\sqrt{1-\beta^2}} \left(\frac{a_x}{\sqrt{1-a^2}} - \frac{\beta}{\sqrt{1-a^2}} \right)$$
$$= \gamma \left(\eta_{ax} - \frac{\beta}{\sqrt{1-a^2}} \right)$$

Momentum and Energy

Here it has been assumed that the relative speed of the observers in Figure 9-1a and b is βc, and $1/\sqrt{1-a^2}$ has been written for η_{oa} in order to make the expression more explicit. Substitution of this and similar expressions for the other particles in Equation 9-12 gives

$$\gamma(m_a\eta_{ax} + m_b\eta_{bx} - m_d\eta_{dx} - m_e\eta_{ex})$$
$$-\beta\gamma\left(\frac{m_a}{\sqrt{1-a^2}} + \frac{m_b}{\sqrt{1-b^2}} - \frac{m_d}{\sqrt{1-d^2}} - \frac{m_e}{\sqrt{1-e^2}}\right) = 0 \quad (9\text{-}13)$$

The expression in the first parenthesis is zero since it just expresses conservation of the x-component of momentum in the unprimed frame of reference. This we have already assumed. However, if momentum is to be conserved in both frames of reference, the expression in the second parenthesis must also be zero. This means that

$$\frac{m_a}{\sqrt{1-a^2}} + \frac{m_b}{\sqrt{1-b^2}} = \frac{m_d}{\sqrt{1-d^2}} + \frac{m_e}{\sqrt{1-e^2}} \quad (9\text{-}14)$$

In other words, the quantity $m/\sqrt{1-u^2}$, summed over all the particles, must be the same before and after the collision, i.e., it must be conserved. This should not be surprising. The requirement arises in exactly the same way that mass conservation did in the non-relativistic case. In fact, for low speeds, the quantity $m/\sqrt{1-u^2}$ reduces to the mass of the particle. For this and other related reasons the quantity $m/\sqrt{1-u^2}$ is often called the "mass" of a moving particle. We shall temporarily denote this "mass" by $\mathfrak{M} = m/\sqrt{1-u^2}$. In this case m would be referred to as the "rest mass" of the particle, the mass it would have when at rest. In this notation the relativistic momentum becomes $p = \mathfrak{M}u$, its classical form, and Equation 9-14 becomes simply the equation for conservation of "mass."

This "mass" of a particle increases with speed, and yet it is a quantity which has no direction associated with it. That is, it has no x-, y-, z-components; it is a scalar, not a vector. In these respects it resembles another classical quantity, the kinetic energy. The analogy is much closer than this, for if we expand the "mass" for cases where the speed is very small (see Appendix A),

$$\mathfrak{M} = \frac{m}{\sqrt{1-u^2}} \approx m + \tfrac{1}{2}mu^2 + \cdots$$

If the equation were to be multiplied by c^2, the second term in the expansion would be the non-relativistic kinetic energy. Classically the first term, mc^2, would be a constant. Since additive constants never enter in calculations involving energy, which always depend on energy differences, the quantity $\mathfrak{M}c^2$ has all the properties of the classical kinetic energy at low speeds.

Multiplying Equation 9-14 by c^2 does not materially affect it. Nevertheless, it states that the quantity $\mathfrak{M}c^2$ is *always* conserved, unlike the classical kinetic energy that is not conserved in inelastic collisions, explosions, and the like. In this respect the quantity $\mathfrak{M}c^2$ behaves like the total energy of an object and we write

$$E = \mathfrak{M}c^2 = \frac{mc^2}{\sqrt{1-u^2}} \tag{9-15}$$

From this point of view, Equation 9-14 becomes simply the conservation of energy equation.

But which is it, a conservation of "mass" equation or a conservation of energy equation? The answer is that it is neither and it is both. The quantity $m/\sqrt{1-u^2}$ is something like the classical mass, but not exactly, and when multiplied by c^2 it becomes $mc^2/\sqrt{1-u^2}$, which behaves like the classical energy, but not exactly. Equation 9-15, that most famous of all equations of relativity, is a peculiar one. It does not state a mathematical relation between two different quantities, energy and "mass." Rather it states that energy and "mass" are *equivalent* concepts. Energy is "mass," and "mass" is energy—except for the constant factor c^2. Equation 9-14 must really be thought of as the conservation equation of some new quantity, neither "mass" nor energy, but having properties of both.

This raises a semantic question. Here is a new quantity that is the relativistic generalization of two classical quantities. Many treatments do just what we have temporarily done and call $\mathfrak{M} =$

Momentum and Energy

$m/\sqrt{1-u^2}$ the "mass," m the rest mass, and call $\mathfrak{M}c^2$ the energy. But, if energy is *always* just c^2 times the "mass," this is not good economy with words or concepts. Furthermore, it seems desirable to have the word *mass* refer to an intrinsic property of the particle, and not to refer to a property of its motion. From now on this book will follow what is becoming quite general practice. The word *energy* will refer to the quantity $mc^2/\sqrt{1-u^2}$. The word *mass* will refer to m, the quantity many other treatments call the "rest mass." In cases where it should cause no confusion, the word *mass* will sometimes be used for the quantity mc^2, the energy of the particle in the frame of reference where it is at rest.

The practice of designating the mass of a particle in energy units has now become very widespread. Almost all tables of the elementary particles now give their masses in MeV (million electron volts), the energy that an electron would gain in being accelerated through a potential difference of 1 million volts. Thus the proton has a mass of 1.673×10^{-27} kg or 938 MeV, arrived at as follows: The mass $m = 1.673 \times 10^{-27}$ kg is equivalent to an energy of $mc^2 = 1.673 \times 10^{-27} \times (2.998 \times 10^8)^2 = 1.504 \times 10^{-10}$ joules. It would therefore be meaningful, but almost never done, to say the mass of a proton is 1.504×10^{-10} joules. Now an electron has a charge of 1.602×10^{-19} coulombs, and when accelerated by 10^6 volts would gain an energy of 1.602×10^{-13} joules. Therefore 1 MeV = 1.602×10^{-13} joules. The mass of a proton, then, is 1.504×10^{-10} joules $\div 1.602 \times 10^{-13}$ joules/MeV, or 938 MeV. This means $mc^2 = 938$ MeV. The convenience of this kind of notation will become clear in succeeding sections.

9-5 AN EXAMPLE

The work of the preceding sections has, no doubt, seemed somewhat formal, and the following discussion of a particular example is in order. Instead of a collision we consider an explosion. Figure 9-3*a* shows two identical objects at rest, each of mass m. The total momentum of the system is zero. Figure 9-3*b* shows the two objects as they fly apart. *If momentum is to be conserved*, i.e., remain zero, the objects must fly apart with velocities uc and $-uc$. To a second observer who

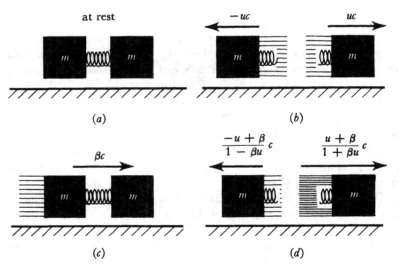

Figure 9-3

is moving toward the left with a velocity $-\beta c$, the explosion appears as in Figure 9-3c and d. Before the explosion the composite object is moving at a velocity βc to the right. After the explosion the two objects fly apart with unequal speeds which may be found from the velocities uc and $-uc$ by using the velocity addition formulas (Equation 5-5). These various speeds are shown in Table 9-1.

Table 9-1. *Velocities of the Objects*

	Frame of reference where explosion occurs from rest (Fig. 9-3a, b)	Frame of reference where original object is moving (Fig. 9-3c, d)
Right-hand object after explosion	uc	$\dfrac{u+\beta}{1+\beta u}c$
Left-hand object after explosion	$-uc$	$\dfrac{-u+\beta}{1-\beta u}c$
Composite object before explosion	0	βc

Momentum and Energy

Table 9-2 is next constructed to show the various momenta. In the frame of reference where the composite object is at rest, its momentum is clearly zero. The momenta of the small objects after the explosion are, from Equation 9-8, $\pm mcu/\sqrt{1-u^2}$.

In the frame of reference where the composite object is moving, its momentum must be $Mc\beta/\sqrt{1-\beta^2}$. Here M has been used for its mass; M *is not* just $2m$, which is the main point to be shown! The momenta of the two smaller objects after the explosion can be obtained most directly from their velocities in Table 9-1. For example, for the right-hand object

$$p = \frac{mc\left(\dfrac{u+\beta}{1+\beta u}\right)}{\sqrt{1-\left(\dfrac{u+\beta}{1+\beta u}\right)^2}} = \frac{mc(u+\beta)}{\sqrt{(1+\beta u)^2 - (u+\beta)^2}}$$

$$= \frac{mc(u+\beta)}{\sqrt{1+\beta^2 u^2 - u^2 - \beta^2}} = \frac{mc(u+\beta)}{\sqrt{1-\beta^2}\sqrt{1-u^2}}$$

The momentum of the left-hand object is similarly obtained. Writing the momenta in terms of the proper velocities and using the Lorentz transformation would have given the same result. The reader is urged to do this.

Table 9-2. *Momenta of the Objects*

	Frame of reference where explosion occurs from rest (Fig. 9-3a, b)	Frame of reference where original object is moving (Fig. 9-3c, d)
Right-hand object after explosion	$\dfrac{mcu}{\sqrt{1-u^2}}$	$\dfrac{mc(u+\beta)}{\sqrt{1-\beta^2}\sqrt{1-u^2}}$
Left-hand object after explosion	$\dfrac{-mcu}{\sqrt{1-u^2}}$	$\dfrac{mc(-u+\beta)}{\sqrt{1-\beta^2}\sqrt{1-u^2}}$
Composite object before explosion	0	$\dfrac{Mc\beta}{\sqrt{1-\beta^2}} = \dfrac{2mc\beta}{\sqrt{1-\beta^2}\sqrt{1-u^2}}$

The total momentum after the explosion, in the frame of reference where the composite object was moving, is the sum of the momenta of the two smaller objects, i.e., the sum of the two upper entries in the right-hand column of Table 9-2. This sum is $2mc\beta/\sqrt{1-\beta^2}\sqrt{1-u^2}$, and *if momentum is to be conserved in this frame of reference too*, this total momentum after the collision must equal the momentum before, as shown in the lower right-hand entry of Table 9-2. It follows immediately that

$$M = \frac{2m}{\sqrt{1-u^2}} \tag{9-16}$$

In other words, the mass of the composite object before the explosion must have been greater than the sum of the masses of its parts! Equation 9-16 is just a special case of Equation 9-14 where the velocity of the mass M is zero.

Thus far, the particular example has shown that if momentum is to be conserved in two frames of reference moving with respect to each other, then masses are not conserved in the collision, but rather the quantity of Equation 9-14, which, multiplied by c^2, we have called the energy. The total energy of the composite object at rest is just Mc^2. The total energy of the two fragments is $2mc^2/\sqrt{1-u^2}$. Equation 9-16 therefore expresses the conservation of relativistic energy:

$$\text{energy before} = Mc^2 = \frac{2mc^2}{\sqrt{1-u^2}} = \text{energy after} \tag{9-17}$$

In case the explosion is not too violent and $u \ll c$, Equation 9-17 can be expanded:

$Mc^2 = 2mc^2 + 2 \times \tfrac{1}{2}mc^2u^2 + \cdots$
$Mc^2 - 2mc^2 = 2 \times \tfrac{1}{2}mc^2u^2$
loss in mass times c^2 = total kinetic energy given to fragments

In low speed cases it appears that the classical kinetic energy comes from the decrease in mass. It would nevertheless be a mistake to consider mass energy on the same level as electrical potential energy, gravitational potential energy, or thermal energy—just as one more storehouse for energy. In the previous example the kinetic energy came from the potential energy stored in the composite object, perhaps from a compressed spring. It *also* "came from" the decrease in mass. The added mass of the composite object *was* the potential energy of the spring.

9-6 MOMENTUM AND ENERGY: A SUMMARY

In our previous discussions the momentum and energy have been written down explicitly in terms of the velocity of a particle of mass m:

$$p_x = \frac{mcu_x}{\sqrt{1-u^2}} \qquad p_y = \frac{mcu_y}{\sqrt{1-u^2}} \qquad p_z = \frac{mcu_z}{\sqrt{1-u^2}}$$

$$p = \frac{mcu}{\sqrt{1-u^2}} \qquad E = \frac{mc^2}{\sqrt{1-u^2}} \tag{9-18}$$

They can be written more compactly in terms of the proper velocity:

$$p_x c = mc^2 \eta_x \qquad p_y c = mc^2 \eta_y \qquad p_z c = mc^2 \eta_z \qquad E = mc^2 \eta_0 \tag{9-19}$$

In doing practical relativistic calculations it is seldom necessary or desirable to use these expressions directly. Later examples will show that it is usually sufficient to use the following easily derived expressions:

$$\mathbf{p}c = \mathbf{u}E \tag{9-20}$$

$$E^2 - p^2 c^2 = m^2 c^4 \tag{9-21}$$

9-7 RELATIVISTIC MOMENTUM AND ENERGY: EXPERIMENT

Calling the quantities in Equation 9-18 the momentum and energy and postulating that they are conserved does not make it true. All that the theoretical work has been able to do is to show that conservation of the quantity called momentum is *consistent* with the postulates of relativity, and that this consistency also requires the conservation of the quantity called energy. The equations also reduce at low speeds to the classical equations, so no violence is done to the wealth of experiments in this domain. But, whether or not these theoretical ideas have anything to do with reality is for experiment to decide.

A. *"Putty Balls."* Many readers will first want to know why such a startling thing as decrease in mass of the fragments of an explosion was not observed as soon as balances were invented. Perhaps the inverse example of a collision will do. Suppose two "putty balls" approach each other at a speed of 10 km/sec and stick together. Now a speed of 10 km/sec is a *very* high one for classical experiments, yet this is only 3×10^{-5} of the speed of light. The change in mass of the resulting agglomeration, because of the thermal energy produced, would be only one part in 10^9. This same thermal energy would raise the temperature of these same objects about 10,000°C and would therefore vaporize them. It is hardly surprising that the equivalence of mass and energy was not discovered in routine investigations of everyday objects!

B. *How Energy Depends on Speed.* A direct test of the energy equation would be to accelerate a particle to a known speed and to slow it down, thus changing its energy into an easily recognizable classical form. Such an experiment was mentioned in Section 5-4, where electrons, whose speeds were directly measured by timing their flight over a known distance, were stopped in a calorimeter where their energy was converted to thermal energy. According to Equation 9-15 electrons with a speed uc would have an energy $E = mc^2/\sqrt{1-u^2}$. After stopping, they would have an energy $E = mc^2$. Each electron would therefore lose an amount of energy $\Delta E = (mc^2/\sqrt{1-u^2}) - mc^2$. In the experiment, the energy of a measured number of electrons was converted to thermal energy. The units on the horizontal axis of Figure 5-4 are not arbitrary as stated there, but are in multiples of mc^2 for each electron. The graph is a graph of u versus $\Delta E = (mc^2/\sqrt{1-u^2} - mc^2)$. The agreement is very good.

C. *"Inelastic" Processes in Nuclear Physics.* In the previous example the electrons remained electrons all through the process. The functional form of Equation 9-15 was checked, but the example said nothing about the startling theoretical conclusion that the mass of a system need not be equal to the sum of the masses of its parts. The first known examples of this are provided by nuclear physics. For instance, the nucleus ^{7}Li is known to be composed of 3 protons and 4 neutrons. A mole (6.0225 × 10^{23} atoms) of ^{7}Li weighs 7.01595 grams. A mole of hydrogen, whose atomic nucleus is a single proton, weighs 1.00782 grams. But a mole of ^{4}He, which is composed of 2 protons and 2 neutrons, weighs 4.00260 grams. These masses are known from mass spectroscopic studies [2] and are directly related to inertial mass and, therefore, to masses obtained by weighing. Therefore, a mole of ^{7}Li plus a mole of hydrogen weigh 0.01857 grams *more* than two moles of helium, which is presumably composed of the same constituents. Therefore, considering the equivalence of mass and energy, we presume that a single atom of hydrogen plus a single atom of ^{7}Li have (.01857/6.0225 × 10^{23}) × (2.9979 × 10^8)^2 = 2.771 × 10^{-12} joules *more* energy than two helium atoms at rest.

If a proton and a ^{7}Li nucleus could be brought together, the resulting combination of 4 protons and 4 neutrons would be expected to break up into two helium nuclei, or α-particles. These α-particles would be expected to have a total kinetic energy of 2.771 × 10^{-12} joules from the change in mass. In a very early nuclear physics experiment, Cockroft and Walton and later Dee and Walton [3] struck stationary ^{7}Li nuclei with protons and showed that sometimes two very energetic α-particles were given off. In a more recent version of the experiment [4] the kinetic energy of the α-particles exceeded the kinetic energy of the incoming proton by 17.344 MeV, or 2.779 × 10^{-12} joules. This value agrees with that calculated from the mass difference within the expected experimental error which is mostly in the mass spectroscopic mass of ^{7}Li.

D. *Particle Disintegrations.* In the physics of nuclei, rather trivial fractions of the total mass are changed in interactions between particles. In the physics of elementary particles really large fractions of the mass are involved. A typical example is that of a K^+ meson which decays into two pions:

$$K^+ \to \pi^+ + \pi^0$$

---

[2] The values given here are calculated from mass spectroscopic doublets given by H. E. Duckworth, B. G. Hogg, and E. M. Pennington, *Rev. Mod. Phys.*, **26**, 463 (1954). The mass scale is based on ^{12}C.

[3] P. I. Dee and E. T. S. Walton, *Proc. Roy. Soc.*, **141**, 733 (1933).

[4] K. F. Famularo and G. C. Phillips, *Phys. Rev.*, **91**, 1195 (1953).

The mass of the K^+ meson is $m_K c^2 = 493.9$ MeV. The mass of the π^+ is $m_+ c^2 = 139.6$ MeV and that of the π^0 is $m_0 c^2 = 135.0$ MeV. All these masses are measured independently—though not, of course, by direct weighing, since K mesons and pions are unstable and only last for a few hundred millionths of a second. A K^+ meson at rest has zero momentum. The pions coming off from the decay, therefore, have zero total momentum and must come off in exactly opposite directions with

$$p_+ + p_0 = 0$$

Conservation of energy gives

$$E_+ + E_0 = m_K c^2$$

Here E_0, E_+, and p_0, p_+ are the energies and the momenta of the neutral and positive pions respectively. Equation 9-21 gives

$$E_+^2 - p_+^2 c^2 = m_+^2 c^4 \qquad E_0^2 - p_0^2 c^2 = m_0^2 c^4$$

Subtracting these equations gives

$$E_+^2 - E_0^2 - p_+^2 c^2 + p_0^2 c^2 = m_+^2 c^4 - m_0^2 c^4$$
$$(E_+ + E_0)(E_+ - E_0) - (p_+ + p_0)(p_+ - p_0)c^2$$
$$= (m_+ + m_0)(m_+ - m_0)c^4$$

Making use of the conservation of momentum and energy equations written down above, we get

$$E_+ - E_0 = \frac{(m_+ + m_0)(m_+ - m_0)c^4}{m_K c^2}$$

$$E_+ + E_0 = m_K c^2$$

Solving these we obtain

$$E_+ = \frac{m_K c^2}{2} + \frac{(m_+ c^2 + m_0 c^2)(m_+ c^2 - m_0 c^2)}{2 m_K c^2}$$

Substituting the mass values gives

$$E_+ = \frac{493.9}{2} + \frac{274.6 \times 4.6}{2 \times 493.9} = 246.9 + 1.3 = 248.2 \text{ MeV}$$

The *kinetic* energy of the π^+ meson is its total energy E_+ minus its rest energy, or mass, or $248.2 - 139.6 = 108.6$ MeV. The kinetic energy of these π^+ mesons coming from K^+ decay has been measured and found to be 107.7 ± 1.0 MeV [5] in satisfactory agreement with the prediction.

If the π^+ and π^0 mesons had had identical masses, then each would have had an energy $m_K c^2/2$ which is to be expected.

None of the previous examples have checked directly the relativistic form of the momentum. The two pions involved in the decay of the K^+ meson have so nearly equal mass that the energy division could have been predicted solely from symmetry. However, much the most surprising results come from the energy considerations just discussed, and these, it should be recalled, followed from the predicted form of the momentum. Other examples more directly involving momentum will be discussed in subsequent chapters.

E. *Gravitational Attraction.* In all our considerations so far, our only concern has been the mechanical interactions of particles in collisions. We have seen that the mass involved in such a collision is completely equivalent to the energy. Is this "mass," which is equivalent in many cases to classical kinetic and potential energy, involved also in gravitational attraction? The answer is that it is. A very famous set of experiments performed by Eötvös, and recently considerably improved by Dicke,[6] show that the mass involved in gravitational attraction is equivalent, i.e., proportional, to the mass involved in collisions. This equivalence has been shown to about 3 parts in 10^{11} for aluminum and gold, and to a lesser accuracy for a variety of substances.

Now the mass of some elements differs from the sum of the masses of their constituent parts by about one percent, so clearly the energies involved in the nuclei of atoms are gravitationally active. Even the potential and kinetic energies of the electrons of atoms make up about one part in 10^5 of the atom's total energy or mass, and this ratio varies considerably from light to heavy atoms; so even such "obviously" classical kinetic and potential energy must contribute to gravitational attraction.

[5] R. W. Birge, *et al.*, *Nuovo Cimento*, **4**, 834 (1956).
[6] R. H. Dicke, "The Eötvös Experiment," *Sci. Am.*, **205**, 84 (1961). P. G. Roll, R. Krotkov, and R. H. Dicke, *Ann. Phys.* (N.Y.), **26**, 442 (1964).

9-8 AN EXAMPLE: A SYMMETRICAL, ELASTIC COLLISION BETWEEN EQUAL MASS PARTICLES

As a particularly simple example of conservation of energy and momentum we consider the collision between an incoming particle of mass m, momentum p_0, and energy E_0, and a stationary particle of equal mass. We consider only the case where the two particles come off with equal energies. The collision is called elastic because the masses of the particles are not altered in the collision. Relativistic energy is conserved in all collisions. Inelastic collisions are simply those in which the masses are altered.

The collision is shown schematically in Figure 9-4. Since we choose only the case where the outgoing particles have equal energy, their momenta are also equal because they have equal mass. Conservation of the y-component of momentum therefore requires that the angles θ made with the incoming direction be equal. Conservation of momentum and energy give

$$p_0 = 2p \cos \theta \tag{9-22}$$
$$E_0 + mc^2 = 2E \tag{9-23}$$

Notice that the rest energy of the struck particle must be included in the energy before the collision. In terms of the energies (from

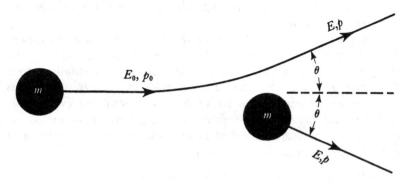

Figure 9-4

Momentum and Energy

Equation 9-21), Equation 9-22 can be written

$$\sqrt{E_0^2 - m^2c^4} = 2\sqrt{E^2 - m^2c^4}\cos\theta$$

Substituting for E from Equation 9-23 gives

$$\sqrt{E_0^2 - m^2c^4} = 2\sqrt{\left(\frac{E_0 + mc^2}{2}\right)^2 - m^2c^4}\cos\theta$$

Solving for $\cos\theta$ gives

$$\cos\theta = \frac{\sqrt{E_0^2 - m^2c^4}}{\sqrt{(E_0 + mc^2)^2 - 4m^2c^4}} = \frac{\sqrt{(E_0 + mc^2)(E_0 - mc^2)}}{\sqrt{(E_0 + 3mc^2)(E_0 - mc^2)}}$$

$$= \sqrt{\frac{E_0 + mc^2}{E_0 + 3mc^2}} \quad (9\text{-}24)$$

It is often convenient to check relativistic results by going to the classical limit of small velocities. At small incoming velocities $E_0 \approx mc^2$. Equation 9-24 then reduces to $\cos\theta = 1/\sqrt{2}$ or $\theta = 45°$. This is the well known classical case. As the total incoming energy becomes very large, i.e., $E_0 \gg mc^2$, $\cos\theta$ approaches 1; θ approaches 0. This is an example of a very general property of relativistic collisions. When the incoming energy becomes much larger than the rest energies involved, all the products of the collision tend to go off very nearly in the direction of the incident particle.

9-9 AN EXAMPLE: A HEAD-ON ELASTIC COLLISION

As a second example of straightforward energy and momentum conservation, we consider a head-on elastic collision in which a mass m, with energy e_0 and momentum p_0, strikes a stationary mass M. In the collision, the mass M is knocked straight forward, with energy E and momentum P, leaving the mass m with e and p.

The algebra needed to find e, p, E, and P is considerably more involved than that of the last example. Careful consideration of all relevant equations in this chapter—particularly look at the example of the previous section—will show that whenever a momentum occurs it is, or may be, multiplied by c, i.e., pc. Whenever a mass occurs it is multiplied by c^2. Considerable effort in writing may be saved and consequent mistakes avoided if wherever pc occurs we simply write p and wherever mc^2 occurs we write m. All quantities then have the dimensions of energy.

Conservation of momentum and energy give

$$p_0 = p + P \tag{9-25}$$
$$e_0 + M = e + E \tag{9-26}$$

Again notice the rest energy of the struck particle is part of the initial energy, here written as M instead of Mc^2. The most direct method of solution of these equations is to write Equation 9-26 in terms of the momenta p and P by using Equation 9-21. Use Equation 9-25 to eliminate p, for example, and solve the resulting equation for P. Thus

$$e_0 + M = \sqrt{p^2 + m^2} + \sqrt{P^2 + M^2}$$
$$(e_0 + M) - \sqrt{P^2 + M^2} = \sqrt{(p_0 - P)^2 + m^2}$$
$$(e_0 + M)^2 + \cancel{P^2} + M^2 - 2(e_0 + M)\sqrt{P^2 + M^2} = p_0^2 - 2p_0 P + \cancel{P^2} + m^2$$

Be on the lookout for expressions of the form $e^2 - p^2$. For instance, on one side $(e_0 + M)^2$ will give an e_0^2 which combined with the p_0^2 from the other side of the equation will give m^2. Thus

$$e_0^2 - p_0^2 + 2e_0 M + M^2 + M^2 - 2(e_0 + M)\sqrt{P^2 + M^2} = -2p_0 P + m^2$$
$$\cancel{m^2} + 2e_0 M + 2M^2 - 2(e_0 + M)\sqrt{P^2 + M^2} = -2p_0 P + \cancel{m^2}$$

Put the radical on one side and square again

$$(e_0 + M)^2(P^2 + M^2) = (e_0 M + M^2 + p_0 P)^2$$

Carry out all the operations to get

$$e_0^2 P^2 + e_0^2 \cancel{M^2} + 2Me_0 P^2 + 2\cancel{M^3 e_0} + M^2 P^2 + \cancel{M^4}$$
$$= e_0^2 \cancel{M^2} + \cancel{M^4} + p_0^2 P^2 + 2\cancel{e_0 M^3} + 2e_0 M p_0 P + 2M^2 p_0 P$$

P now comes out as a factor. This is encouraging since $P = 0$ must be a solution to the problem, i.e., the incoming mass misses, leaving

Momentum and Energy

the mass M at rest! Then we solve for P obtaining

$$P = \frac{2p_0 M(e_0 + M)}{2Me_0 + M^2 + m^2} \tag{9-27}$$

where we have again set $e_0^2 - p_0^2 = m^2$. Substituting into Equation 9-25 we get

$$p = \frac{p_0(m^2 - M^2)}{2Me_0 + M^2 + m^2} \tag{9-28}$$

There is no doubt that the algebra of relativistic mechanics is more complicated than that of classical mechanics. A large part of the remaining chapters will be devoted to the development of ideas that lead to simpler algebra. There are a few characteristics of exercises such as this which are worth mentioning. Expressions get very long, but judicious use of $e^2 - p^2 = m^2$ will usually simplify things at each stage. In successive squarings to get rid of radicals, very high powers of the variable sought will be developed. These usually cancel. If the c's are left out, every parameter and variable has the dimensions of an energy; therefore every term must contain the same number of factors.

In the classical limit $e_0 \approx m$ and Equations 9-27 and 9-28 become

$$P = p_0 \frac{2M}{m + M} \qquad p = p_0 \frac{m - M}{m + M}$$

These may be familiar to many readers; they are, in any case, correct.

One special case is worthy of note. If $m = M$, Equations 9-27 and 9-28 reduce to $P = p_0$, $p = 0$. Thus the incoming particle stops, and the struck particle goes on with the entire momentum, just as in the classical case. This should serve as a warning against taking the idea of a mass that increases with speed too literally in a classical sense. For if the incoming particle were very energetic, its "mass" would be larger than the mass about to be struck. It is a familiar classical result (roll a basketball into a ping-pong ball) that when a heavy mass strikes a lighter one, both go forward. That is not the case here.

9-10 ENERGY AND MOMENTUM IN TWO DIFFERENT FRAMES OF REFERENCE

Equation 9-18 gives the momentum and energy of a particle whose velocity is uc. From the point of view of an observer moving past that frame of reference, the particle has a different velocity and hence a different momentum and energy. If the momentum and energy of the particle are written in terms of the proper velocity as in Equation 9-19, Equation 8-6 can be used to find the relation between momenta and energies in two different systems. Thus

$$\begin{aligned} p_x{}'c &= \gamma(p_x c - \beta E) \\ E' &= \gamma(E - \beta p_x c) \\ p_y{}'c &= p_y c \qquad p_z{}'c = p_z c \end{aligned} \qquad (9\text{-}29)$$

These can be solved for p_x, p_y, and E obtaining

$$\begin{aligned} p_x c &= \gamma(p_x{}'c + \beta E') \\ E &= \gamma(E' + \beta p_x{}'c) \\ p_y c &= p_y{}'c \qquad p_z c = p_z{}'c \end{aligned} \qquad (9\text{-}30)$$

Therefore, momentum and energy in two frames of reference are related by the same Lorentz transformation that relates the coordinates and time of an event in these frames of reference.

Another useful and interesting relation can be derived from the last of Equation 9-21. The mass of a particle is its mass and remains so in all frames of reference. (A handbook which gives the mass of an electron does not change its printing because it moves rapidly!) Therefore we may write

$$E'^2 - p'^2 c^2 = m^2 c^4 = E^2 - p^2 c^2 \qquad (9\text{-}31)$$

This can also be derived directly from Equation 9-29 or 9-30. The reader is urged to do this.

Exercises

1. Derive Equations 9-20 and 9-21 from Equation 9-18.
2. We neglected to show that if energy is conserved in one inertial system, it is conserved in all. Take Equation 9-14, which expresses conservation of

Momentum and Energy

energy (if multiplied by c^2), but express it in the primed system, i.e., write a'^2 instead of a^2. Next express the velocities a', in terms of the a's and β, by means of the velocity addition formulas. Manipulate this, following the pattern used for proper velocities in Chapter 8. Show finally that energy is conserved in one system if energy and momentum are conserved in the other.

3. The weights of a mole of each of these substances are:

^{2}H (deuterium): 2.014102 grams
^{6}Li (lithium-six): 6.015126 grams
^{4}He (helium-four): 4.002603 grams

In the reaction ^{2}H + ^{6}Li → 2^{4}He, the incoming deuteron has negligible momentum and energy.

a. By how much does the mass of a mole of deuterium plus a mole of lithium-six exceed the mass of two moles of helium into which it turned?

b. How much mass is "changed" to kinetic energy in a single reaction if a mole contains 6.025×10^{23} atoms?

c. How many MeV does each helium nucleus have after the reaction? (1 MeV = 1.6×10^{-13} joules)

d. How many kilowatt-hours can the reaction of a mole of deuterium with a mole of lithium supply?

4. A π^+ meson has a mass or rest energy of 139.6 MeV. If mesons of total energy 174.5 MeV are produced in a high energy accelerator, how far will they travel before half of them have decayed? Take the π^+ meson half-life to be 1.77×10^{-8} sec. If the mesons have an energy of 13,960 MeV how far will they travel before half of them decay?

5. *a.* If a proton of *kinetic* energy 437 MeV collides elastically with a proton at rest and the two protons rebound with equal energies, what is the included angle between them? (R. B. Sutton, *et al.*, *Phys. Rev.*, **97**, 783 (1955), find 84.0° ± 0.2° for the experimental result.)

b. If the incoming proton has a total energy of 33,000 MeV (the most energetic now produced in any accelerator), what is the included angle between them?

6. Restore the c's in Equations 9-27 and 9-28.

7. A mass M traveling at a speed $4c/5$ makes a completely inelastic collision with an identical mass, i.e., they stick together. With what speed does the resulting particle travel? What is its mass, i.e., its energy in a system where it is at rest?

8. A K^+ meson at rest sometimes breaks up into three charged π mesons. All charged pions have the same mass.

a. If all three pions go off with equal energies, what energy does each get? What angles do the pions make with each other?

b. What is the maximum energy one of the three pions can get? This occurs when the other two go off together in the opposite direction.

c. What is the least energy one of the three pions can get? What energies do the other two have in this case?

d. Can a K^+ meson decay into four pions?

9. A K^+ meson traveling through the laboratory breaks up into two π mesons. One of the π mesons is left at rest. What was the energy of the K^+? What is the energy of the remaining π meson?

10. In one particular frame of reference an object has an energy E^* and zero momentum. Use Equation 9-29 to find its energy and momentum in any other frame of reference moving with velocity uc with respect to the first. How do these results compare with Equation 9-18?

11. A moving object approaches a stationary one with a velocity uc. At what velocity βc would an observer have to move so that in his frame of reference the objects would have equal and opposite velocities? Use the velocity addition formulas.

12. If the two objects in the preceding problem have equal masses m such that the momentum of the incoming object is p_0, using Equation 9-29 find β for that frame of reference where the momenta of the two objects are equal and opposite; compare to the result of the previous problem. What is the total energy of the two objects in this frame of reference?

Answers to Completion Sentences on Page 130

a. $\dfrac{mcv}{\sqrt{1-v^2}}$

b. $u \cos \theta$ as before

c. $\dfrac{u \sin \theta}{\sqrt{1-u^2 \cos^2 \theta}}$ as before

d. $p \sin \theta = \dfrac{mc \dfrac{u \sin \theta}{\sqrt{1-u^2 \cos^2 \theta}}}{\sqrt{1 - \dfrac{u^2 \sin^2 \theta}{1-u^2 \cos^2 \theta}}} = \dfrac{mcu \sin \theta}{\sqrt{1-u^2}}$

hence $p = \dfrac{mcu}{\sqrt{1-u^2}}$

10

Particles of Zero Mass

10-1 LIGHT FLASHES AS "PARTICLES"

ANYONE WHO HAS BASKED in the warmth of the summer sun knows that light carries energy. More sober experiments show that light can push on an object; light therefore carries momentum. Maxwell's electromagnetic theory predicts that a flash of light containing an energy E should have a momentum E/c. This prediction has been well verified by experiment. Light therefore has well defined mechanical properties. Does the relativistic mechanics of the last chapter in any way apply to flashes of light?

Equations 9-18, 9-20, and 9-21 summarize the relevant mechanical properties of a particle of mass m:

$$pc = \frac{mc^2 u}{\sqrt{1-u^2}} \qquad E = \frac{mc^2}{\sqrt{1-u^2}} \qquad (9\text{-}18)$$

$$pc = uE \qquad (9\text{-}20)$$

$$E^2 - p^2 c^2 = m^2 c^4 \qquad (9\text{-}21)$$

If a light flash is to be treated as a particle, then it clearly must move "with the speed of light." Therefore $u = 1$ and Equation 9-20 becomes

$$pc = E \qquad (10\text{-}1)$$

If Equation 10-1 is inserted in Equation 9-21, it shows that $m = 0$. The momentum and energy of this light flash are still finite since the zero in the numerator is "canceled" by the zero in the denominator. We therefore conclude that it is perfectly consistent with our relativistic mechanics to treat a light flash as a particle of zero mass. It is very gratifying to find that the special theory of relativity which was born to reconcile conflicts in the kinematical properties of light and ordinary objects, also includes their mechanical properties in a single all inclusive system.

10-2 PHOTONS

That a flash of light behaves mechanically like a particle of zero mass follows directly from relativistic mechanics. It is not surprising, then, that it was Einstein, acting on Planck's interpretation of radiation from hot bodies, who predicted that light would have particle properties, and that the energy and the frequency of the light particle, or *quantum* as it was called, should be proportional.

That the energy and frequency of the quantum might be proportional can be seen from a consideration of the Doppler effect and the Lorentz transformation of energy. Suppose a source emits a flash of light of frequency f and this light is observed by an observer moving directly away from the source with a speed βc. He observes a frequency f', where f' is given from Equation 5-14 as

$$f' = f\sqrt{\frac{1-\beta}{1+\beta}} \qquad (10\text{-}2)$$

Suppose that the source emits a light flash of total energy E. The observer moving away sees in his frame of reference an energy E', given by Equation 9-29 as

$$E' = \gamma(E - \beta p_x c)$$

Particles of Zero Mass

Since this is a light flash going in the x direction, $p_x c = E$ and

$$E' = \frac{1}{\sqrt{1-\beta^2}} E(1-\beta) = E\sqrt{\frac{1-\beta}{1+\beta}} \tag{10-3}$$

Equations 10-2 and 10-3 are of exactly the same form. A physical law that says that the energy of a quantum of light is proportional to its frequency is, therefore, a relativistically acceptable one, i.e., it would be the same for all observers in inertial frames. Einstein proposed that the energy and frequency of a light quantum are, in fact, related by

$$E = hf \tag{10-4}$$

where h is a universal constant called Planck's constant. This prediction has been completely verified, and h has been found experimentally to have the value 6.6256×10^{-34} joule/sec, or 4.1356×10^{-21} MeV/sec.

When the particle nature of light is to be emphasized, the particle of light is called a *photon*, a name that has largely superceded the original, *quantum*.

10-3 OTHER PARTICLES OF ZERO MASS

In the radioactive decay of nuclei a particle called a *neutrino* is given off. These are also emitted in the decay of certain elementary particles, though there is recent evidence that in some decays these neutrinos are not identical to those of nuclear radioactivity. As nearly as experiment has been able to determine, neutrinos are also massless and therefore have their momentum and energy related like Equation 10-1.

Although it has never been observed experimentally, physicists think that there is probably another massless particle, the *graviton*, associated with gravity in the same way the photon, or light, is associated with electricity and magnetism.

These are all the known particles whose masses are thought to be really zero. However, it is sometimes convenient to treat a particle whose energy is very much greater than its mass, as massless. Equation 9-21 reduces to Equation 10-1 if E and pc are both much greater than mc^2.

10-4 AN EXAMPLE: AN ATOM ABSORBS LIGHT

A hydrogen atom sitting still in one particular frame of reference absorbs a photon of wavelength 1215.7×10^{-10} meters. With what speed does the resulting excited atom recoil?

Light of wavelength 1215.7×10^{-10} meters has a frequency $f = c/\lambda = 2.46 \times 10^{15}$ cycles/sec. A photon of light of this frequency has an energy $E_0 = hf = 6.63 \times 10^{-34} \times 2.46 \times 10^{15} = 1.63 \times 10^{-18}$ joules $= 10.2 \times 10^{-6}$ MeV. Suppose M is the mass of the hydrogen atom before it absorbs the photon. Suppose P is the momentum of the hydrogen atom that has absorbed the photon, and E is its energy. Since the atom is initially at rest, the entire momentum before the "collision" is just the momentum of the photon: E_0/c. The total energy is $E_0 + mc^2$. Conservation of momentum and energy give

$$Pc = E_0 \qquad E = E_0 + Mc^2$$

For any particle, Equation 9-20 gives $Pc = uE$. For the recoiling hydrogen atom, where $Pc = E_0$ and $E = E_0 + Mc^2$, we have

$$u = \frac{E_0}{Mc^2 + E_0} \tag{10-5}$$

The recoiling particle, therefore, travels at this speed. Since $E_0 = 10.2 \times 10^{-6}$ MeV and $Mc^2 = 938$ MeV, E_0 may be neglected in the denominator and the speed of the particle becomes

$$uc = 3 \times 10^8 \times \frac{10.2 \times 10^{-6}}{938} = 3.26 \text{ m/sec}$$

a very small speed for an atomic sized particle!

The mass, M^*, of the recoiling particle can be found very easily from Equation 9-21

$$M^{*2}c^4 = E^2 - P^2c^2 = (E_0 + Mc^2)^2 - E_0^2 = M^2c^4 + 2E_0Mc^2$$

$$M^*c^2 = Mc^2\sqrt{1 + \frac{2E_0}{Mc^2}} \tag{10-6}$$

Particles of Zero Mass

The mass of the resulting particle is greater than that of the original hydrogen atom as we might have expected. Since $E_0 \ll Mc^2$ we may expand the radical obtaining

$$M^*c^2 \approx Mc^2 + E_0 \tag{10-7}$$

In this case, the mass of the excited atom is greater than the mass of the absorbing atom by the energy of the absorbed photon. Since $E_0 \ll Mc^2$, the fractional difference is not large.

This process is important astronomically. Absorption of light by hydrogen atoms in space drives them away from stars, thus counteracting the gravitational attraction of the star for interstellar hydrogen.

10-5 AN EXAMPLE: DECAY OF THE K^+ MESON

In Section 9-7D the decay of a K^+ meson at rest into a positively charged and a neutral π meson was discussed. The energy of the π^+ meson was derived. Similarly the energy of the π^0 meson can be calculated:

$$E_0 = m_K c^2 - E_+ = \frac{m_K c^2}{2} - \frac{(m_+ c^2 + m_0 c^2)(m_+ c^2 - m_0 c^2)}{2 m_K c^2}$$

The π^0 meson of energy E_0 subsequently decays into two photons. We wish to calculate the maximum and minimum energy the two photons can have in the laboratory. The maximum and minimum energy will be obtained when the photons travel in the same direction as, and directly opposite to, the motion of the π^0 meson as shown in Figure 10-1c.

The two photons have energies E_1 and E_2 and momenta E_1/c and $-E_2/c$, where use has been made of the zero-mass relation $pc = E$. Conservation of energy and momentum give

energy: $E_1 + E_2 = E_0$
momentum: $p_0 c = \sqrt{E_0^2 - m_0^2 c^4} = E_1 - E_2$

Solving for E_1 and E_2 gives

$$E_1 = \frac{E_0}{2} + \frac{1}{2}\sqrt{E_0^2 - m_0^2 c^4} \qquad E_2 = \frac{E_0}{2} - \frac{1}{2}\sqrt{E_0^2 - m_0^2 c^4}$$

Reference to Section 9-7D will show $E_0 = 245.7$ MeV, $m_0 c^2 = 135.0$ MeV.

$$E_1 = \frac{245.7}{2} + \frac{1}{2}\sqrt{245.7^2 - 135.0^2} = 122.8 + 102.8$$
$$= 225.6 \text{ MeV}$$
$$E_2 = 122.8 - 102.8 = 20.0 \text{ MeV}$$

Another possible case is shown in Figure 10-1d where the π^0 meson moving through the laboratory decays into two photons of equal energy, E_3, making equal angles θ with the direction of travel of the π^0 meson. What is the angle θ?

Conservation of energy gives $E_3 =$ _____ (a). Since the photons are massless, they each have a momentum $p_3 c =$ _____ (b). The x-component of this momentum is therefore $p_{3x}c =$ _____ (c), and must be related by conservation of momentum

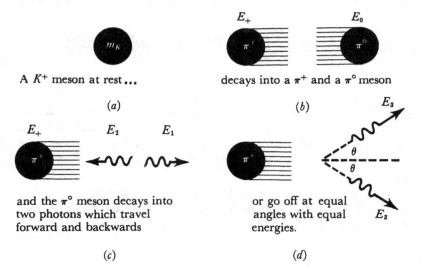

A K^+ meson at rest... decays into a π^+ and a π^0 meson

(a) (b)

and the π^0 meson decays into two photons which travel forward and backwards

or go off at equal angles with equal energies.

(c) (d)

Figure 10-1

Particles of Zero Mass

to the original momentum of the π^0 meson $p_{3x}c = $ _____ (d). Solving for cos θ, we get, cos $\theta = $ _____ (e). Putting in actual numbers, this makes $\theta = $ _____ (f). (Answers are at the end of the chapter.)

Actually photons come off from the moving π^0 mesons in all directions between the two extreme cases treated here.

Exercises

1. A π^+ meson decays at rest into a μ meson and a neutrino. What is the energy of the μ meson? What is its "kinetic energy," i.e., its total energy minus its mass?

2. A K^+ meson at rest decays into a π^+ meson and a π^0 meson. The π^+ meson decays into a μ meson and a neutrino. What is the maximum energy of the final μ meson? What is its minimum energy?

3. The π^0 meson from the decay of a K^+ meson at rest has an energy of 245.7 MeV.

a. With what speed is this meson moving through the laboratory?

b. It sends off a flash of light at right angles to its direction of motion *as measured in the system of the π^0 meson.* Use the velocity addition formulas to find its direction with respect to the direction of travel of the π^0 meson in the laboratory.

c. How does this compare to answer (f) of Section 10-5?

4. A K^+ meson frequently decays into a μ^+ meson and a neutrino. If a K^+ meson decays at rest, what is the energy of the μ^+? the neutrino? The value found experimentally for the μ^+ kinetic energy [R. W. Birge, et al., *Nuovo Cimento* **4**, 834 (1956)] is 152.36 ± 1.1 MeV.

5. Treat a "light flash" as a particle of momentum p and energy E, and use the Lorentz transformation on momentum and energy to derive the aberration formula, Equation 5-7. Remember, the direction of a light flash is the direction of its momentum.

6. At very high energy accelerators these days one of the most important fields of research is "neutrino physics." Neutrinos are neutral and cannot be bent or focussed in magnets. It is therefore very difficult to prepare high intensity beams of neutrinos. What is actually done is to make a very high intensity pion beam. The pions decay into a muon and a neutrino. (See Exercise 1.) Suppose the pions have a total energy of 1396 MeV, i.e., a gamma of ten. What is the maximum angle the decay neutrino can make with the original pion beam? Do the neutrinos stay pretty well "focussed"?

7. Equation 10-7 is only approximate. Why?

Answers to Completion Sentences on Pages 156–157

a. $E_3 = \dfrac{E_0}{2}$

b. $p_3 c = E_3 = \dfrac{E_0}{2}$

c. $p_{3x} c = \dfrac{E_0}{2} \cos \theta$

d. $\dfrac{E_0 \cos \theta}{2} = \dfrac{p_0 c}{2} = \dfrac{1}{2} \sqrt{E_0^2 - m_0^2 c^4}$

e. $\cos \theta = \dfrac{\sqrt{E_0^2 - m_0^2 c^4}}{E_0} = \sqrt{1 - \left(\dfrac{m_0 c^2}{E_0}\right)^2}$

f. $\cos \theta = \sqrt{1 - \left(\dfrac{135}{245.7}\right)^2} = 0.836 \qquad \theta = 33.3°$

11

Center-of-Mass and Particle Systems

11-1 WHEN IS AN OBJECT AT REST?

IN CONSIDERING MANY ATOMIC COLLISION PROCESSES we think of an atom as a single object and often refer to it as a "particle." At other times we consider it as a complex system composed of a nucleus surrounded by many moving electrons. Considered as a single particle it is more or less obvious what we mean when we say an atom is "at rest." Considered as a complex system, none of whose parts are "at rest," we need a more precise definition. Such a complex system is considered "at rest" when the total momentum of the system is zero. In classical physics a frame of reference where the total momentum of a group of particles is zero is called the center-of-mass frame, because it is the frame of reference where the center-of-mass is at rest. Relativistically it is hard to define the center-of-mass of a group of particles. (Is it the center-of-mass or center-of-energy? Should you say that the kinetic energy of a particle is at the position of the particle? Whose time? etc.) However, the frame of reference where the total momentum is zero is easily defined. Physicists sometimes call this the center-of-momentum frame, but the common practice is to take over the classical term center-of-mass, as we shall do.

11-2 TOTAL MOMENTUM AND ENERGY OF A GROUP OF PARTICLES

Suppose the ith single particle in a group of particles has momentum p_{xi}, p_{yi} and energy e_i in one frame of reference. In a second frame of reference it will have momentum and energy given by Equation 9-29

$$p_{xi}'c = \gamma(p_{xi}c - \beta e_i)$$
$$e_i' = \gamma(e_i - \beta p_{xi}c)$$
$$p_{yi}'c = p_{yi}c$$

The total momentum P and energy E of the group of particles will be the sum of the momenta and energies of the single particles of the group

$$P_x'c = \sum_i p_{xi}'c = \gamma\left(\sum_i p_{xi}c - \beta \sum_i e_i\right)$$
$$= \gamma(P_x c - \beta E) \tag{11-1}$$

And similarly

$$E' = \gamma(E - \beta P_x c) \tag{11-1'}$$
$$P_y'c = P_y c \tag{11-1''}$$

In other words, the relation between *total* momentum and energy in two frames of reference is given by the same Lorentz transformation that governs those of a single particle. From Equation 11-1 it is easy to show that

$$E^2 - P^2 c^2 = E'^2 - P'^2 c^2 \tag{11-2}$$

11-3 THE CENTER-OF-MASS FRAME OF REFERENCE

For any group of particles it is possible to find a frame of reference in which the total momentum is zero.[1] In this center-of-mass frame we will denote the total momentum and energy by P^* and E^*. Of course, $P^* = 0$. Since there is no y-component of momentum in the center-of-mass frame, there will be no y-component anywhere. We therefore drop the subscript x and write from Equation 11-1 and 11-2

$$P^*c = 0 = \gamma(Pc - \beta E)$$
$$E^* = \gamma(E - \beta Pc)$$
$$E^{*2} - P^{*2}c^2 = E^{*2} = E^2 - P^2c^2$$

From the momentum equation we get $Pc = \beta E$. Substituting this in the energy equation we get $E^* = \sqrt{1 - \beta^2}\, E$. Summarizing, then:

$$\beta = \frac{Pc}{E} \quad E = \frac{E^*}{\sqrt{1 - \beta^2}} \quad Pc = \frac{E^*\beta}{\sqrt{1 - \beta^2}} \quad (11\text{-}3)$$
$$E^2 - P^2c^2 = E^{*2}$$

Since βc is the velocity with which the center-of-mass frame is moving with respect to the frame of reference of P and E, βc may very well be said to be the velocity of the group of particles considered as a whole. Equation 11-3 has exactly the form that it would for a single particle (see Equations 9-18, 9-20, and 9-21) provided that $E^* = mc^2$. The energy in the center-of-mass plays the role of the mass of a single particle. But, of course, mc^2 is also the energy of a single particle in its center-of-mass! Relativistically all distinction is lost between a group of particles and a single particle!

[1] The single exception to this statement is the case of a group of zero mass particles all going the same direction.

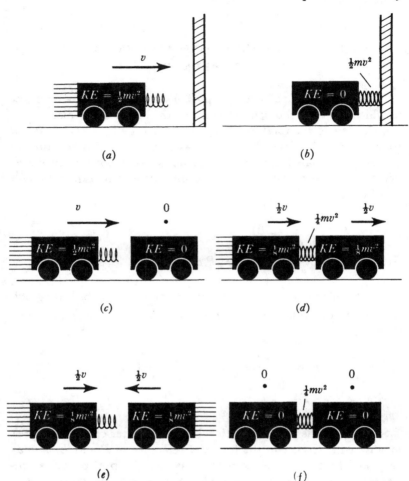

Figure 11-1. (a) A cart moving with velocity v has a kinetic energy $\frac{1}{2}mv^2$. (b) When it hits a solid wall it slows down and stops before rebounding. The kinetic energy is zero. All the energy, $\frac{1}{2}mv^2$, is stored in the spring. (c) But when a cart with the same energy, $\frac{1}{2}mv^2$, hits a stationary cart of equal mass, conservation of momentum requires (d) that at the maximum compression of the spring, the carts retain a kinetic energy of $\frac{1}{4}mv^2$, leaving only $\frac{1}{4}mv^2$ to compress the spring. (e) In the center-of-mass system each cart has a velocity $\frac{1}{2}v$ and kinetic energy $\frac{1}{8}mv^2$. Since the total momentum is zero in the center-of-mass system, (f) no kinetic energy is retained, and the entire energy, $\frac{1}{4}mv^2$, is available to compress the spring.

11-4 IMPORTANCE OF ENERGY IN THE CENTER-OF-MASS

In any physical process, it takes energy in the center-of-mass to do anything. If an automobile going 61 mi/hr creeps up behind and bumps an automobile going 59 mi/hr, nothing much happens—provided, of course, the drivers do not lose control of the cars. The total kinetic energy of the two cars is quite sufficient to cause a catastrophe, as witness the results in a head-on collision at the same speeds. But in the slight bump at high speeds the amount of energy *available* for destructive purposes is, fortunately, quite small.

A more academic situation is provided by a cart of velocity v carrying a spring as in Figure 11-1a. Since, for now, only the familiar classical limit of low velocities is to be considered, the kinetic energy of the cart is $\frac{1}{2}mv^2$. When the cart collides with a very massive, fixed wall, the spring compresses until the entire kinetic energy of the cart, $\frac{1}{2}mv^2$, is stored as potential energy of the spring. At this point, the cart is momentarily at rest before it bounces back. See Figure 11-1b. Contrast this with the case where the same cart with the same energy collides with another identical cart at rest, as in Figure 11-1c. As the carts collide, the spring compresses, but the force of the compressed spring drives the second cart away. At the instant when the spring is at its maximum compression, the two carts are traveling at the same speed. *Conservation of momentum* requires that the speed of both carts at this instant be $v/2$, i.e., $mv = m(v/2) + m(v/2)$. The kinetic energy of both carts is $\frac{1}{2}m(v/2)^2 + \frac{1}{2}m(v/2)^2 = \frac{1}{4}mv^2$. The difference between this and the original kinetic energy is $\frac{1}{2}mv^2 - \frac{1}{4}mv^2 = \frac{1}{4}mv^2$. Therefore the energy available to compress the spring is only one half the total.

Only a fraction of the total energy was available to do work in the preceding example because momentum conservation required that some of that energy be retained as kinetic energy of the whole system at all stages in the collision. There is only one frame of reference where no kinetic energy need be retained by the entire system, and that is obviously the frame of reference where the momentum is zero, the center-of-mass frame. In the preceding classical example the center-of-mass frame moves at a speed $v/2$ to the right. In this frame of reference the cart with the spring is moving at a speed $v/2$ to the right, the struck cart at a speed $v/2$ to the left. The total momentum is therefore zero, as required. The kinetic energy of each cart is therefore $\frac{1}{2}m(v/2)^2$ in this frame of reference. The total energy available to compress the spring is therefore $2 \times \frac{1}{2}m(v/2)^2 = \frac{1}{4}mv^2$ as before. The energy available to compress the spring is the total energy in the center-of-mass system!

The same argument goes over into relativistic mechanics. The total energy available to *do* anything is the total energy of the center-of-mass system E^*.

11-5 AN EXAMPLE: A COLLISION BETWEEN EQUAL MASS PARTICLES

To illustrate the principles of the previous sections, we treat a collision between equal mass particles. This is of great importance in nuclear physics where most generators of high energy particles accelerate protons to high energies and these protons then collide with either protons or neutrons at rest, or nearly at rest. When two equal mass protons collide at very high energies, several things can happen. The most obvious of these is that the two protons bounce apart like billiard balls. This is called elastic scattering. The kinetic energy of the incoming proton can be changed into π mesons, K mesons, or other particles. Finally the kinetic energy of the incoming proton can be changed into a proton-antiproton pair. We will also consider this process.

A. *Elastic Scattering of Equal Mass Particles.* Figure 11-2a shows a diagram of a particle of mass m, energy E_a, and momentum p_a approaching a stationary particle of identical mass m. The incoming particle is deflected in the collision by an angle θ_d, and leaves with a momentum p_d and energy E_d. The struck particle has initial momentum, and energy $p_b = 0$ and $E_b = mc^2$. After the collision it recoils with momentum p_e and energy E_e at an angle θ_e. Since

Center-of-Mass and Particle Systems

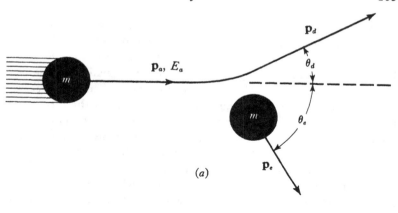

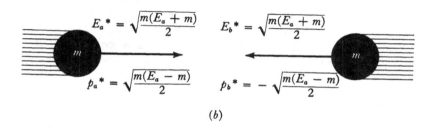

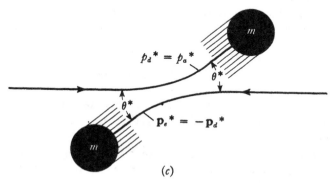

Figure 11-2. (a) *Laboratory system collision.* (b) *Before the collision in the center-of-mass frame, the particles approach with equal energies.* (c) *After the collision the particles fly off in opposite directions in the center-of-mass frame.*

the collision is elastic, the particles both leave with mass m. In Section 9-8 momentum and energy conservation were used directly to solve this problem for the symmetrical case $\theta_d = \theta_e$, $E_d = E_e$. In this discussion, scattering through any angle will be considered by the use of the center-of-mass reference frame. The reader should understand from the outset that this could be done without reference to the center-of-mass by momentum and energy conservation in the original reference frame alone. In fact, the reader should try it. However, not only is the algebra considerably simplified by the use of the center-of-mass frame, but in advanced work this frame of reference is often the convenient one in which to answer theoretical questions about how *often* two protons scatter by various angles—a question to be avoided here.

The algebra is sufficiently complicated that we again use the shorthand notation $pc \rightarrow p$, $mc^2 \rightarrow m$, where all quantities have the units of energy. The total momentum of this system is p_a; its total energy $E_a + m$. Any system with this momentum and energy, be it a single particle or system of particles, moves as a whole with a velocity

$$\beta = \frac{p_a}{E_a + m} \tag{11-4}$$

(See Equation 11-3.) In the case of our system of particles, this is to be interpreted as the speed of the center-of-mass reference frame. In the center-of-mass reference frame the whole system has an energy E^* given by

$$\begin{aligned} E^{*2} &= E^2 - p^2 = (E_a + m)^2 - p_a^2 \\ &= E_a^2 - p_a^2 + 2mE_a + m^2 = 2m(E_a + m) \end{aligned} \tag{11-5}$$

Again the energy in the center-of-mass of the system has been treated just as if the system were a single particle and this energy in the center-of-mass were its mass.

An alternative way of treating the problem, once β has been found, is to find the momenta and energies of the individual particles by means of a Lorentz transformation. Denoting quantities in the center-of-mass system by * we get

Center-of-Mass and Particle Systems

$$p_{ax}{}^* = \gamma(p_a - \beta E_a) = \frac{1}{\sqrt{1 - \dfrac{p_a{}^2}{(E_a + m)^2}}} \left(p_a - \frac{p_a}{E_a + m} E_a \right)$$

$$= \frac{E_a + m}{\sqrt{2m(E_a + m)}} \frac{m}{E_a + m} p_a$$

$$= \sqrt{\frac{m}{2(E_a + m)}}\, p_a = \sqrt{\frac{m}{2(E_a + m)}} \sqrt{E_a{}^2 - m^2}$$

$$= \sqrt{\frac{m(E_a - m)}{2}} \tag{11-6}$$

$$p_{bx}{}^* = \gamma(p_b - \beta E_b) = \gamma(0 - \beta m) = -\frac{E_a + m}{\sqrt{2m(E_a + m)}} \frac{p_a m}{E_a + m}$$

$$= -\sqrt{\frac{m}{2(E_a + m)}}\, p_a = -p_{ax}{}^* \tag{11-7}$$

This merely verifies by direct calculation that the particles have equal and opposite momenta in the center-of-mass system as shown in Figure 11-2b. The energies may also be obtained by a Lorentz transformation:

$$E_a{}^* = \gamma(E_a - \beta p_a) = \frac{E_a + m}{\sqrt{2m(E_a + m)}} \left(E_a - \frac{p_a{}^2}{E_a + m} \right)$$

$$= \frac{E_a + m}{\sqrt{2m(E_a + m)}} \frac{m(E_a + m)}{(E_a + m)} = \sqrt{\frac{m(E_a + m)}{2}} \tag{11-8}$$

$$E_b{}^* = \gamma(E_b - \beta p_b) = \frac{E_a + m}{\sqrt{2m(E_a + m)}} (m - \beta 0) = \sqrt{\frac{m(E_a + m)}{2}} \tag{11-9}$$

The particles have the same energy in the center-of-mass system, as expected, and the sum of their energies is just E^* calculated earlier. In fact, these Lorentz transformations were quite superfluous for doing the problem, since symmetry tells us immediately that each particle will have an energy $E^*/2$ obtainable from Equation 11-5. The algebra is much more tedious.

In the center-of-mass system the particles collide and are each deflected by an angle θ^* as shown in Figure 11-2c. The two particles must go off in opposite directions after the collision in order that momentum will remain zero, i.e., be conserved. Since energy is conserved, the total energy of the system in the center-of-mass frame is still E^*. Since the collision is elastic, the masses are still m, the momenta are equal, and E^* is equally divided between the two particles. In short, after the collision, the energies are unaltered, and the momenta are only altered in direction. Therefore

$$p_{dx}^* = \cos\theta^* \sqrt{\frac{m(E_a - m)}{2}} \qquad p_{ex}^* = -\cos\theta^* \sqrt{\frac{m(E_a - m)}{2}}$$

$$p_{dy}^* = \sin\theta^* \sqrt{\frac{m(E_a - m)}{2}} \qquad p_{ey}^* = -\sin\theta^* \sqrt{\frac{m(E_a - m)}{2}}$$

$$E_d^* = \sqrt{\frac{m(E_a + m)}{2}} \qquad E_e^* = \sqrt{\frac{m(E_a + m)}{2}}$$

A Lorentz transformation back to the laboratory system gives the momenta and energies there. For example,

$$p_{dx} = \gamma(p_{dx}^* + \beta E_d^*) = \frac{E_a + m}{\sqrt{2m(E_a + m)}}$$
$$\left(\cos\theta^* \sqrt{\frac{m(E_a - m)}{2}} + \frac{p_a}{E_a + m}\sqrt{\frac{m(E_a + m)}{2}}\right)$$

$$= \frac{E_a + m}{\sqrt{2m(E_a + m)}}$$
$$\left(\cos\theta^* \sqrt{\frac{m(E_a - m)}{2}} + \frac{\sqrt{E_a^2 - m^2}}{E_a + m}\sqrt{\frac{m(E_a + m)}{2}}\right)$$

$$= \frac{\sqrt{E_a^2 - m^2}}{2}(1 + \cos\theta^*)$$

$$p_{dy} = p_{dy}^* = \sin\theta^* \sqrt{\frac{m(E_a - m)}{2}}$$

Center-of-Mass and Particle Systems

from which

$$\tan \theta_d = \frac{p_{dy}}{p_{dx}} = \sqrt{\frac{2m}{E_a + m}} \frac{\sin \theta^*}{1 + \cos \theta^*} = \sqrt{\frac{2m}{E_a + m}} \tan \frac{\theta^*}{2} \quad (11\text{-}10)$$

Similarly,

$$\tan \theta_e = \sqrt{\frac{2m}{E_a + m}} \tan \frac{\pi - \theta^*}{2} \quad (11\text{-}11)$$

If the incoming particle is moving slowly, $E_a \approx m$ and the radicals in Equations 11-10 and 11-11 reduce to one. Then $\theta_d = (\theta^*/2)$ and $\theta_e = (\pi/2) - (\theta^*/2)$; $\theta_d + \theta_e = \pi/2$. This is the familiar result that in an elastic collision between equal mass particles, the particles go off at right angles to each other. As the incoming energy E_a gets much larger than m, the angles θ_d and θ_e get smaller and smaller. This is a typical relativistic effect. At very high energies, the products of any kind of collision go very nearly in the forward direction, and this effect depends on the ratio of the incoming energy to the mass of the interacting particles. The details differ, of course, according to the particular kind of collision being considered. The solution of this problem for E_d and E_e is left to the student (Exercise 2). The angles and energies of the final products of the collision are still given in terms of the center-of-mass angle θ^*. In principle, a complete solution gives E_d in terms of θ_d. If given θ_d, one must think of solving Equation 11-10 for θ^* and using this value of θ^* to compute E_d.

B. *Production of Antiprotons.* When a very high energy proton hits another at rest, several things can happen other than elastic scattering. Other particles can be produced, one of the most interesting of which is the antiproton. It is a particle of identical mass to the proton, but with a negative charge. When produced, it must always be produced with another proton, not just by itself. It will not decay if left to itself, but will annihilate with a proton if it hits one; both proton and antiproton disappear and, usually, about five π mesons are produced. Our question is, how much energy must be given to a proton so that when it collides with a second proton at rest, a proton and antiproton can be produced?

The problem is especially simple when solved in the center-of-mass frame of reference (Figure 11-3). In this frame, two protons approach with equal energies E_a^* and $E_b^* = E_a^*$. The *least* energy in the center-of-mass that will produce a proton and antiproton is that energy which will leave all four particles sitting still, i.e., the total energy in the center-of-mass is $4m$. In order to get a center-of-mass energy $E^* = 4m$, it is necessary to have a laboratory energy E_a given by Equation 11-5 as

$$E^{*2} = (4m)^2 = 2m(E_a + m) \qquad E_a = 7m$$

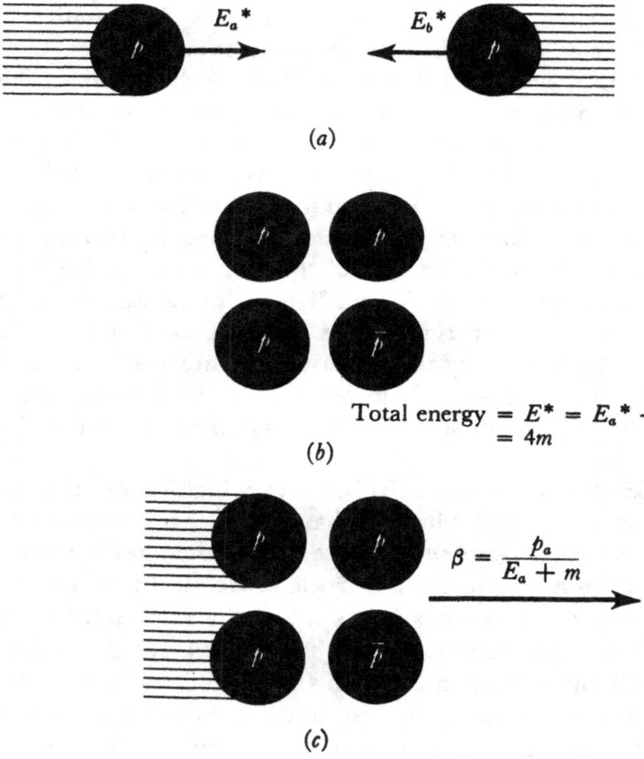

Figure 11-3. (a) *In the center-of-mass frame, two protons approach each other with equal energies* (b) *and transform their energy into a new proton and an antiproton at rest in the center-of-mass frame.* (c) *But meanwhile, back at the laboratory, the four particles are moving with the speed of the center-of-mass frame.*

Center-of-Mass and Particle Systems

Of this, m is provided by nature in the rest energy of the proton. $6m = 5628$ MeV must be provided by the accelerator. A proton must be accelerated by means of a voltage of 5.6 billion volts. Several such accelerators exist. In fact, the Bevatron at the University of California was carefully designed to have an energy of 6.6 billion volts in order to make antiprotons—if they existed—and leave some energy over to do experiments with them. Chamberlain and Segré discovered the existence of the antiproton with this accelerator and a very cleverly designed experiment for which they were awarded the Nobel prize.

Exercises

1. Show that for small velocities, Equations 11-4 and 11-5 reduce to the classically expected expressions discussed in Section 11-4.

2. In the elastic scattering of two equal masses, discussed in Section 11-5A, find the final energies of the two particles E_d and E_e in terms of θ^*. Eliminate θ^* between your expression and Equations 11-10 and 11-11 to get E_d in terms of θ_d.

3. Show that Equations 11-10 and 11-11 are consistent with Equation 9-24.

4. Certain physicists have proposed that a magnetic monopole exists, i.e., an isolated "north" magnetic pole. There are reasons to believe that it might have a mass 137^2 times the electron mass. If magnetic monopoles must be produced in pairs—a north and a south pole together—how high an energy proton would need to collide with a proton at rest in order to produce a pair of monopoles? If such a monopole pair were just barely produced, so as to have zero kinetic energy in the center-of-mass, what would be the energy of each in the laboratory?

5. If a particle of mass m and energy E_0 collides with a stationary identical particle, find the velocity of the center-of-mass frame of reference, Equation 11-4, by use of the velocity addition formula and the requirement that in the center-of-mass frame the particles have oppositely directed velocities of equal magnitude.

6. What is the minimum energy of an electron that will produce a π^0 meson in a collision with another electron at rest? Mesons may be created singly; they do not have to be created in pairs. After the production, the electrons remain.

7. The masses of a normal helium nucleus ^{4}He, a proton ^{1}H, and a triton ^{3}H are given below. If a photon of energy 300 MeV in the laboratory hits a ^{4}He nucleus so as to break it into a proton and a triton, which travel

off at right angles to the photon direction in the center-of-mass system, find (a) the proton kinetic energy and angle in the laboratory and (b) the triton kinetic energy and angle. mc^2 (of ^{4}He) = 3728 MeV, $mc^2(^1\text{H})$ = 938 MeV, $mc^2(^3\text{H})$ = 2809 MeV.

8. If two photons of energy E_1 and E_2 *approach* each other in the laboratory, what is the speed of the center-of-mass frame with respect to the laboratory? If the two photons are going the same direction can you find an answer?

9. Suppose one photon has an energy of 200 MeV and is traveling along the x-axis. Suppose another has an energy of 100 MeV and is traveling along the y-axis. What is the total energy of this system? The total momentum? If a single particle had this same total energy and momentum, what would be its mass? In what direction would it be traveling? With what speed?

10. If a particle of mass m and energy E_0 hits a stationary particle of mass M, what is the center-of-mass energy of the system?

11. A pion hits a proton at rest to produce a K meson and a lambda particle. What is the minimum pion energy for which this can occur? The masses of the particles are: pion, 139.6 MeV; proton, 938.2 MeV; K meson, 497.8 MeV (this is a neutral K meson which has a mass slightly greater than the charged K); lambda, 1115.4 MeV.

12. A problem often faced by a high energy nuclear physicist is the following: Three pions (assume the mass 140 MeV) are seen to leave the same point in a nuclear interaction. They have energies and make angles with each other as shown on the diagram. Assuming they came from the decay of some single particle, find the mass of that particle. Was it a K meson, mass 500 MeV? In what direction was it moving?

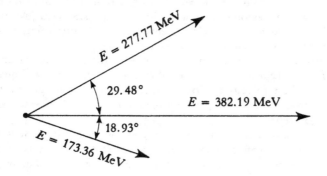

Center-of-Mass and Particle Systems

13. A long-lived neutral K meson (mass 497.8 MeV) is not supposed to decay into two pions. Yet, in a very recent experiment, in a beam of these neutral K mesons two pions were seen to emerge from a single point in the beam. One had a momentum of 656 MeV and made an angle of 18.32° to the direction of the beam. The other had a momentum of 940 MeV and an angle of 12.36°. They were in the same plane. Calculate the mass of the particle that decayed into two pions. Calculate its direction to the beam. What do you think? Was it a long-lived K meson that decayed? [A. Abashian, et al., Phys. Rev. Letters **13**, 243 (1964).]

14. Figure 11-4a shows a very famous bubble chamber picture; Figure 11-4b shows a diagram of the significant tracks; Table I gives the momenta and angles of the various particles. Azimuthal angles are measured clockwise in the plane of the picture from an arbitrary axis roughly straight up and down the page; dip angles are the angles the tracks make with the plane of the paper. The momenta are directly measured from the curvature of the tracks except for 7 and 8, which are photons, and whose momenta are obtained from the electron pairs produced. Work in x, y, z coordinates and work accurately since errors accumulate. All data are taken from V. E. Barnes, et al., Phys. Rev. Letters **12**, 204 (1964).

a. Assume the photons 7 and 8 came from the decay of a single particle. What are the momentum, energy, and mass of the particle? What kind of particle is it likely to be?

b. Assume track 5 is a π^- meson and track 6 a proton. The experimenters believe this from measuring the bubble densities of the tracks; they also tried other assumptions which just didn't fit. What are the momentum,

Table 11-1. *Measured Quantities*

Track	Azimuth (deg)	Dip (deg)	Momentum (MeV/c)
1	4.2 ± 0.1	1.1 ± 0.1	4890 ± 100
2	6.9 ± 0.1	3.3 ± 0.1	501 ± 5.5
3	14.5 ± 0.5	−1.5 ± 0.6	...
4	79.5 ± 0.1	−2.7 ± 0.1	281 ± 6
5	344.5 ± 0.1	−12.0 ± 0.2	256 ± 3
6	9.6 ± 0.1	−2.5 ± 0.1	1500 ± 15
7	357.0 ± 0.3	3.9 ± 0.4	82 ± 2
8	63.3 ± 0.3	−2.4 ± 0.2	177 ± 2

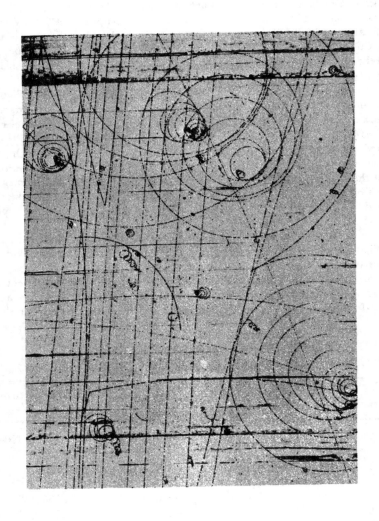

Figure 11-4a

Center-of-Mass and Particle Systems

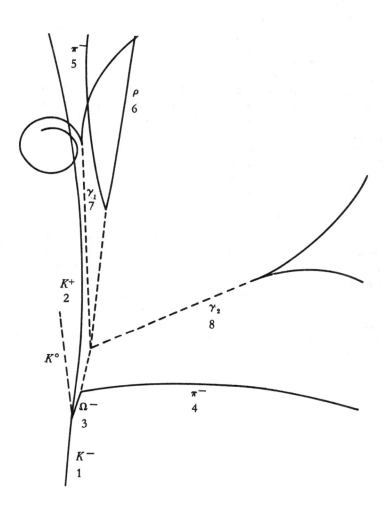

Figure 11-4b

energy, and mass of the particle which decayed into 5 and 6? What is it likely to be?

  c. Assuming that a single particle decayed into the products found in (a) and (b) what are the momentum, energy, and mass of this particle? Is it likely to be a known particle?

  d. Assume track 3 decayed into track 4, a π^- meson, and the particle you found in (c). What are the momentum, energy, and mass of this particle? How does your calculated direction check with the measured one? Remember, the errors may be considerable. The momentum of track 3 was not directly measured because the track is too short for a reliable curvature measurement. If you found a mass near 1680 MeV, you have just "discovered" the Ω^- particle that had been predicted to decay into a Ξ^0 and π^- and to have a mass of 1680 MeV. This bubble chamber picture was the one on which it was discovered.

  e. The Ω^- was predicted to be produced in the reaction

$$K^- + p \to K^0 + K^+ + \Omega^-$$

Track 1 is the incoming K^- which hits a stationary proton to produce the reaction; track 2 is an outgoing K^+. Assuming your momenta and energies in (d) are correct, what is the "missing" momentum and energy of the neutral particle produced in the reaction? Is it likely to be a K^0 meson? If it is, this checks the production process for the Ω^-.

  f. Go back to part (b). Suppose, instead, that track 6 were a pion. What would be the mass of the decaying particle? Is this the mass of a known neutral particle, within errors? (This technique, you see, is not without its problems.)

12

Four-Vectors

12-1 THE ENERGY-MOMENTUM FOUR-VECTOR

SUPPOSE YOU ARE A PHYSICIST out on a space platform (this is to eliminate gravity), and you have just finished a series of collision experiments on a plane table which tells you that the x- and y-components of momentum are conserved. You believe, of course, that space is isotropic, there is no preferred direction in it, and, therefore, that all frames of reference rotated in any way are equivalent to each other. You therefore conclude that you would have obtained the same results had your table been rotated 90°, making the new axis the old z-axis. Since you believe that no experiment prefers any set of axes, you conclude that the new y-component of momentum will be conserved. Therefore, the old z-component of momentum will be certain to be conserved—even though you had not measured it.

The fact that space is isotropic allowed, in fact required, the extension of information obtained from measurements in two dimensions into a third dimension. The isotropy of space is the principal reason that vectors are useful physical concepts.

Isotropy of space means that no particular orientation of a coordinate system is to be preferred over any other. The principle of relativity extends this isotropy of space to include uniformly moving coordinate systems: No uniformly moving frame of reference is to be preferred over any other. Motion involves time, and it is in this way that relativity extends the symmetry of space into a fourth dimension of time.

The requirement that relativistic energy be conserved arose in exactly the same way that the requirement of conservation of the z-component of momentum arose for the space physicist of the previous example. In Chapter 9 conservation of momentum was postulated. Relativity required the law of conservation of momentum to hold in all uniformly moving systems, just as the isotropy of space required that the x- and y-momentum be conserved in all rotated coordinate systems. This symmetry among uniformly moving systems led directly to the requirement of the conservation of a fourth quantity to accompany the three components of the momentum. This fourth conserved quantity was identified as the energy. Since it arises out of space-time symmetries in exactly the same way as conservation of the third component of the momentum vector arises from two dimensions and rotational symmetry of space, the energy is often called the fourth component of a four dimensional momentum-energy vector, or four-vector.

12-2 THE LORENTZ TRANSFORMATION AS A ROTATION IN FOUR DIMENSIONS

Since ordinary three dimensional space is isotropic, it cannot make any difference what rotated coordinate system is used to describe the components of a vector quantity. Figure 12-1a shows a vector **A** in a (x,y) coordinate system. In this coordinate system **A** is said to have the components A_x and A_y. In Figure 12-1b the *same* vector **A** is shown in a (x',y') coordinate system rotated with respect to the first by the angle φ. It has a component A_x' in this coordinate system represented by $\overline{OQ}$. Now $A_x' = \overline{OQ} = \overline{OS} + \overline{ST} + \overline{TQ}$. But, $\overline{OS} = \overline{OR}\cos\varphi$; $\overline{ST} = \overline{RT}\sin\varphi$; $\overline{TQ} = \overline{TP}\sin\varphi$. Thus $A_x' = \overline{OR}\cos\varphi + (\overline{RT} + \overline{TP})\sin\varphi$. But $\overline{OR} = A_x$, and $\overline{RT} + \overline{TP} = \overline{RP} = A_y$. Thus, with a similar expression for A_y'

$$A_x' = \cos\varphi\, A_x + \sin\varphi\, A_y$$
$$A_y' = -\sin\varphi\, A_x + \cos\varphi\, A_y \qquad (12\text{-}1)$$

Four-Vectors

If the vector **A** is described in terms of its components in two completely equivalent coordinate systems, Equation 12-1 states the relation the components must bear to one another. In fact, matters are sometimes reversed and a vector is formally defined as a quantity described by two (or three in a three dimensional space) numbers that transform into one another like those of Equation 12-1.

The motion of a particle can be described in terms of its momentum and energy in two completely equivalent inertial systems moving with respect to one another. The relationship between the momentum and energy in those two systems is stated by the Lorentz transformation, with the trivial y- and z-components ignored.

$$p_x'c = \gamma p_x c - \beta \gamma E$$
$$E' = -\beta \gamma p_x c + \gamma E \qquad (12\text{-}2)$$

If we write P_x for $p_x c$, and P_0 for E to emphasize that momentum and energy are really components of the same physical quantity

$$P_x' = \gamma P_x - \beta \gamma P_0$$
$$P_0' = -\beta \gamma P_x + \gamma P_0 \qquad (12\text{-}3)$$

If $\gamma \to \cos \varphi$ and $\beta \gamma \to \sin \varphi$, then Equations 12-3 and 12-1 bear a striking resemblance to each other.

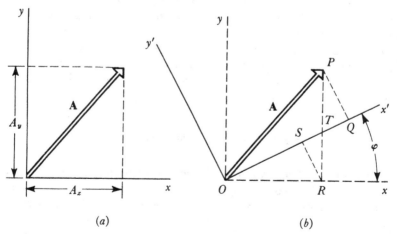

Figure 12-1

If we take one more formal step and write instead of $P_0 = E$, $P_4 = iE$ ($i = \sqrt{-1}$), then Equation 12-3 becomes

$$P_x' = \gamma P_x + i\beta\gamma P_4$$
$$P_4' = -i\beta\gamma P_x + \gamma P_4 \tag{12-4}$$

With the identification $\gamma \to \cos\varphi$ and $i\beta\gamma \to \sin\varphi$, the identity of Equations 12-4 and 12-1 becomes complete since $\gamma^2 + (i\beta\gamma)^2 = 1$ exactly as $\cos^2\varphi + \sin^2\varphi = 1$.

Similarly the Lorentz transformation of the coordinates of an event may be written

$$x' = \gamma x + i\beta\gamma(ict)$$
$$ict' = -i\beta\gamma x + \gamma(ict) \tag{12-5}$$

The Lorentz transformation may therefore be thought of as a rotation in a four dimensional space-time, where the fourth dimension is ict, and rotations are by imaginary angles. Any quantity which transforms from one moving coordinate system to another according to the Lorentz transformation is called a four-vector. Thus momentum-energy, proper velocity, and the space and time interval between events are four-vectors.

The mathematics of rotations and rotational symmetries is a very highly developed subject, and once the Lorentz transformation has been identified as a rotation, all these powerful mathematical tools may be applied to relativity. It is unlikely that most beginning readers would find this much of a help. Subsequent derivations, therefore, may be suggested by vector rotations, but will be proved directly from the Lorentz transformations.

12-3 ORDINARY SCALARS

Scalars are physical quantities which may be described by a single number that does not change its value when the coordinate system is rotated. Temperature at a point is one example; the mass of an object is another. The x-component of a vector is *not* a scalar; it may be expressed by a single number, but it changes when the coordinate system is rotated. Another example of a scalar is the *length of a*

Four-Vectors

vector. If the vector **A** in Figure 12-1 is 3 inches, it remains 3 inches no matter how the coordinate system may be rotated.

The length A of a vector **A** may be expressed in terms of its components. Thus the square of the length of **A**, written A^2, is written in the coordinate system of Figure 12-1a as $A^2 = A_x^2 + A_y^2$; in that of Figure 12-1b, it would be $A^2 = A_x'^2 + A_y'^2$. Since the length is the same in both we have

$$A^2 = A_x'^2 + A_y'^2 = A_x^2 + A_y^2 \tag{12-6}$$

Although the truth of Equation 12-6 is evident geometrically, it can be demonstrated formally from Equation 12-1. Thus

$$\begin{aligned} A^2 = A_x'^2 + A_y'^2 &= (\cos \varphi A_x + \sin \varphi A_y)^2 \\ &\quad + (-\sin \varphi A_x + \cos \varphi A_y)^2 \\ &= (\cos^2 \varphi + \sin^2 \varphi) A_y^2 + (\sin^2 \varphi + \cos^2 \varphi) A_y^2 \\ &\quad + (2 \cos \varphi \sin \varphi - 2 \sin \varphi \cos \varphi) A_x A_y \\ &= A_x^2 + A_y^2 \end{aligned}$$

The sum of the squares of the components of a vector, therefore, forms a scalar quantity called the length of the vector.

Another useful scalar formed from two vectors is the dot- or scalar-product. In Figure 12-2 the vectors **A** and **B** add to give **C**

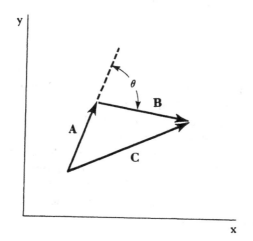

Figure 12-2

in the familiar way. The vector addition $\mathbf{C} = \mathbf{A} + \mathbf{B}$ can be expressed in terms of components, thus

$$C_x = A_x + B_x$$
$$C_y = A_y + B_y \tag{12-7}$$

Then

$$\begin{aligned} C^2 &= C_x{}^2 + C_y{}^2 = (A_x + B_x)^2 + (A_y + B_y)^2 \\ &= A_x{}^2 + A_y{}^2 + B_x{}^2 + B_y{}^2 + 2(A_xB_x + A_yB_y) \\ &= A^2 + B^2 + 2(A_xB_x + A_yB_y) \end{aligned}$$

From the law of cosines

$$C^2 = A^2 + B^2 + 2AB \cos \theta$$

Therefore

$$AB \cos \theta = A_xB_x + A_yB_y \equiv \mathbf{A} \cdot \mathbf{B} \tag{12-8}$$

The quantity $AB \cos \theta$ is clearly a scalar since it is the (length of $\mathbf{A}$) $\times$ (length of $\mathbf{B}$) $\times$ (cosine of the angle between $\mathbf{A}$ and $\mathbf{B}$), none of which change with rotation of the coordinate system. This quantity is called the scalar product of the vectors $\mathbf{A}$ and $\mathbf{B}$ and denoted by $\mathbf{A} \cdot \mathbf{B}$.

12-4 FOUR-SCALARS OR LORENTZ INVARIANTS

Just as a scalar in ordinary three dimensional space is a quantity which does not change when the coordinate system is rotated, so a *four-scalar* is a quantity which remains the same in two frames of reference moving with respect to each other. Four-scalars are more often called Lorentz-invariants, since they do not change when transformed according to a Lorentz transformation.

The most familiar example of a four-scalar is the mass of a particle. The mass of an electron is the same no matter how fast the laboratory is moving. The square of the mass of a particle is given according to Equation 9-21 as

$$m^2c^4 = E^2 - p_x{}^2c^2 - p_y{}^2c^2 - p_z{}^2c^2$$

Four-Vectors

or if we write $mc^2 = M$, $E = P_0$, $p_x c = P_x$, etc., we have

$$M^2 = P_0^2 - P_x^2 - P_y^2 - P_z^2 \tag{12-9}$$

Equation 12-9 is completely analogous to Equation 12-6—except for the change in sign which may also be understood by the use of i as before. The mass of a particle is sometimes referred to as the length of its momentum.

It is clear from its definition that the mass of a particle is a Lorentz invariant, but it can be formally demonstrated in a way analogous to the demonstration that the length of an ordinary vector is invariant in rotated coordinate systems. Thus, writing Equation 12-9 in the primed coordinate system

$$\begin{aligned} M^2 &= P_0'^2 - P_x'^2 - P_y'^2 - P_z'^2 \\ &= (\gamma P_0 - \beta\gamma P_x)^2 - (\gamma P_x - \beta\gamma P_0)^2 - P_y^2 - P_z^2 \\ &= (\gamma^2 - \beta^2\gamma^2)P_0^2 - (\gamma^2 - \beta^2\gamma^2)P_x^2 - P_y^2 - P_z^2 \\ &= P_0^2 - P_x^2 - P_y^2 - P_z^2 \end{aligned} \tag{12-10}$$

The same proof obviously holds for any quantity that transforms with a Lorentz transformation, i.e., any four-vector. The square of the fourth component minus the sum of the squares of the three spacial components is a quantity that remains the same in any inertial frame of reference.

The four-vector analogy to Equation 12-8 is this: If **A** and **B** are two four-vectors (bold face sans-serif type denotes a four-vector) then their scalar product is written

$$\mathbf{AB} = A_0 B_0 - A_x B_x - A_y B_y - A_z B_z \tag{12-11}$$

The proof that this is a Lorentz invariant is left to the reader (Exercise 1).

12-5 FOUR-VECTORS

Relativity considers the four elements, or components, which make up a four-vector to make up a single physical quantity. This is clearest in the case of the four-vector interval between events. There is even a name for it: interval. It is easy to think of "the interval" as taking care of spacial and temporal intervals in one single four dimensional entity. Furthermore, "the interval" is nearly always expressed directly in terms of its four components: the x-, y-, and z-components of the displacement between the events and the time interval between them. In talking about a single quantity it is natural to use the same units for each component, hence the use of $x_0 = ct$ as the fourth component.

The momentum-energy is also considered relativistically to be a single quantity with four components. Curiously it is usually called the momentum four-vector, but its components are given in energy units: $P_x = p_x c, \ldots P_0 = E$. Just as an ordinary vector can be specified in terms of its three components, or alternatively in terms of its length and two angles, the four-momentum may be specified by four quantities chosen in different ways. For example, the three components of the momentum and the energy, i.e., the four components, may be given. The mass of the moving particle may then be determined. More usually the mass is given along with the energy and direction, again four quantities.

The three spacial components of a four-vector always form an ordinary vector in three dimensional space. Ordinary vectors are not necessarily the first three components of a four-vector. Thus the four-momentum has momentum, a vector, for its spacial part. On the other hand, velocity is a perfectly good ordinary vector, but does not transform with a Lorentz transformation and is not a four-vector. In the next chapter we shall see that electric field is another ordinary vector that does not form part of a four-vector.

The fourth- or time-component of a four-vector is a scalar in three dimensional space. Thus energy and the time interval between events are scalars. Four-scalars, such as mass, are also scalars in three dimensional space.

A four-vector differs from an ordinary vector in one important respect. The length of a four-vector can be zero and yet the four-vector can have non-zero components. The length of a photon's four-momentum, i.e., its mass, is zero, and yet the photon can have momentum and energy!

Four-Vectors

12-6 AN EXAMPLE: ELASTIC SCATTERING OF EQUAL MASS PARTICLES

In order to contrast the use of four-vector techniques to those used earlier, we will apply them to the elastic scattering of equal mass particles, the example of Section 11-5A.

Figure 12-3a shows this collision in the laboratory system where a

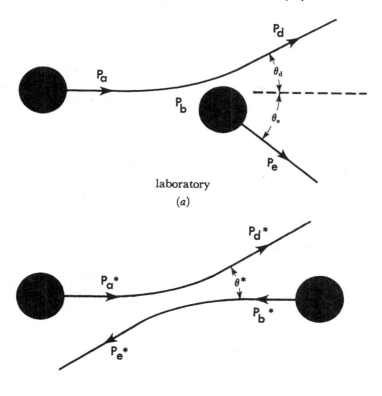

laboratory

(a)

center-of-mass

(b)

Figure 12-3

Table 12-1. Laboratory Four-Momenta

	x-component	y-component	t-component
P_a	$\sqrt{e_a^2 - m^2}$	0	e_a
P_b	0	0	m
P_d	$\cos\theta_d \sqrt{e_d^2 - m^2}$	$\sin\theta_d \sqrt{e_d^2 - m^2}$	e_d
P_e	$\cos\theta_e \sqrt{e_e^2 - m^2}$	$-\sin\theta_e \sqrt{e_e^2 - m^2}$	e_e

particle of four-momentum P_a strikes a stationary particle of four-momentum P_b. The two outgoing particles have four-momenta P_d and P_e. Since the incoming particle of mass m is incident in the x-direction with energy e_a, it has momentum $P_{ax} = \sqrt{e_a^2 - m^2}$, $P_{ay} = 0$. Of course $P_{a0} = e_a$. Since it is stationary, the struck particle has $P_{bx} = P_{by} = 0$ and $P_{b0} = m$. The outgoing particles have momenta and energies given in Table 12-1.

The problem could now be solved directly in the laboratory system by straightforward application of conservation of momentum and energy—or, in our new language, by conservation of all four components of the momentum four-vector. However, as in Section 11-5, it is most convenient to solve the problem in the center-of-mass system. In this system the incoming particles have equal and opposite momenta, and since they have equal mass, they have equal energy. Thus $P_{ax}^* = -P_{bx}^* = \sqrt{e^{*2} - m^2}$, where $P_{a0}^* = P_{b0}^* = e^*$, the energy of each in the center-of-mass. These are entered in Table 12-2. In the center-of-mass system the conservation equations

Table 12-2. Center-of-Mass Four-Momenta

	x-component	y-component	t-component
P_a^*	$\sqrt{e^{*2} - m^2}$	0	e^*
P_b^*	$-\sqrt{e^{*2} - m^2}$	0	e^*
P_d^*	$\cos\theta^* \sqrt{e^{*2} - m^2}$	$\sin\theta \sqrt{e^{*2} - m^2}$	e^*
P_e^*	$-\cos\theta^* \sqrt{e^{*2} - m^2}$	$-\sin\theta \sqrt{e^{*2} - m^2}$	e^*

Four-Vectors

can be satisfied without direct application by noting the symmetry of the problem. The particles enter with equal and opposite momenta and equal energies; they leave, after having been deflected by the angle θ^*, with equal and opposite momenta; since their masses are unchanged, their final energies are equal to the initial ones, e^*.

The problem is solved in the center-of-mass system. It only remains to express the solution in the laboratory system. In Section 11-5 the Lorentz transformation was used for this. Here we use four-vector invariants. The scalar product of two four-vectors is an invariant. Therefore in order to find e^* in terms of e_a we form the product

$$\mathsf{P}_a \mathsf{P}_b = \mathsf{P}_a{}^* \mathsf{P}_b{}^*$$

or

$$P_{a0}P_{b0} - P_{ax}P_{bx} - P_{ay}P_{by} = P_{a0}{}^*P_{b0}{}^* - P_{ax}{}^*P_{bx}{}^* - P_{ay}{}^*P_{by}{}^*$$

Substituting values from Tables 12-1 and 12-2

$$e_a m = e^* e^* + \sqrt{e^{*2} - m^2}\, \sqrt{e^{*2} - m^2}$$

or

$$e^{*2} = \frac{m}{2}(e_a + m) \tag{12-12}$$

Compare Equation 12-12 with 11-8. Similarly, to find one of the outgoing energies, e_d, in terms of e^* we form another product

$$\mathsf{P}_b \mathsf{P}_d = \mathsf{P}_b{}^* \mathsf{P}_d{}^*$$
$$P_{b0}P_{d0} - P_{bx}P_{dx} - P_{by}P_{dy} = P_{b0}{}^*P_{d0}{}^* - P_{bx}{}^*P_{dx}{}^* - P_{by}{}^*P_{dy}{}^*$$

Again substituting from Tables 12-1 and 12-2

$$m e_d = e^* e^* + \sqrt{e^{*2} - m^2}\, \sqrt{e^{*2} - m^2}\, \cos \theta^*$$
$$m e_d - m^2 = (e^{*2} - m^2)(1 + \cos \theta^*)$$

Substituting for e^* from Equation 12-12

$$m(e_d - m) = \left(\frac{m e_a}{2} + \frac{m^2}{2} - m^2\right)(1 + \cos \theta^*)$$
$$= m(e_a - m)\frac{(1 + \cos \theta^*)}{2}$$

Finally,

$$(e_d - m) = (e_a - m) \cos^2 \frac{\theta^*}{2} \tag{12-13}$$

Equation 12-13 gives the energy e_d in terms of the scattering angle in the center-of-mass. The four-vector product $\mathbf{P}_a \mathbf{P}_d = e_a e_d - p_a p_d \cos \theta_d$ involves $\cos \theta_d$ and may be used to get θ_d in terms of center-of-mass angles.

Exercises

1. Prove Equation 12-11.
2. In the example of Section 12-5 take the four-momentum product $\mathbf{P}_d \mathbf{P}_e = \mathbf{P}_d{}^* \mathbf{P}_e{}^*$. Prove that for slowly moving particles where $e \approx m$, the angle between the outgoing particles is 90°.
3. A particle of energy e_a and mass m collides with a stationary particle of equal mass, and the two stick together. Suppose the incoming particle has four-momentum $\mathbf{P}_a$, the struck particle $\mathbf{P}_b$, and the composite particle $\mathbf{P}_d$. Consider the problem in the center-of-mass where two particles of energy e^* collide to form a particle of mass $M = 2e^*$ at rest. By considering the invariant $\mathbf{P}_a \mathbf{P}_b$, find e^* in terms of e_a. By considering $\mathbf{P}_b \mathbf{P}_d$, find the energy of the composite mass in the laboratory.
4. Show that the proper time between two events is the "length" of the four-vector interval between them. What does it mean if the square of the "length" is negative?
5. A π^0 meson of mass m and total energy E decays, in the laboratory, into two photons, one of which makes an angle θ^* with the direction of travel of the meson. θ^* is measured in the center-of-mass system. Find the angle between the two photons in the laboratory by four-vector methods.

13

Electric and Magnetic Fields and Forces

13-1 ELECTRIC AND MAGNETIC FIELDS BETWEEN PLANE SHEETS OF CHARGE

ELECTRIC AND MAGNETIC FIELDS ARE, in a relativistic sense, different aspects of the same phenomenon. In Figure 13-1a a charged parallel plate condenser lies at rest. The electric field between the plates of width w and length l is

$$E_y = \frac{Q}{\epsilon_0 l w} \tag{13-1}$$

where Q is the charge on the condenser. The MKS system of units is used here. The electric field is in the y-direction.

In a frame of reference O' moving toward positive x with a speed βc, the charges are moving to the left with a speed βc. Moving charges constitute a current, and the currents produce a magnetic field between the plates. It is the same condenser in both cases. Whether the field is purely electric or has a magnetic field associated with it depends on the frame of reference from which it is viewed.

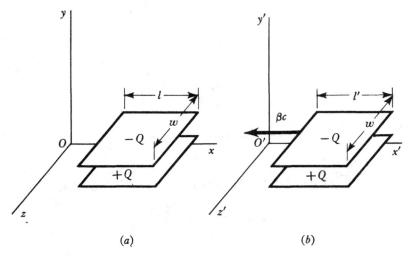

Figure 13-1

The magnetic field between parallel ribbons of width w carrying equal currents I in opposite directions is $B = \mu_0 I/w$. The current is the total charge that passes a point per unit time. In Figure 13-1b the condenser moves its whole length in a time $l'/\beta c = l\sqrt{1-\beta^2}/\beta c$. The current flow in each plate is therefore $I = Q\beta c/l\sqrt{1-\beta^2}$. The magnetic field is toward negative z and has a magnitude

$$B_z' = -\frac{\mu_0 Q \beta c}{lw\sqrt{1-\beta^2}} = -\gamma\epsilon_0\mu_0\beta c E_y \tag{13-2}$$

The magnetic field has been expressed in terms of the electric field in the system where the condenser is at rest.

Not only does a magnetic field appear in the frame of reference where the condenser moves, but the electric field is increased because the length of the plates contracts and leaves the same charge distributed on a smaller area. The electric field is given by the same expression as Equation 13-1:

$$E_y' = \frac{Q}{\epsilon_0 l'w} = \frac{Q}{\epsilon_0 lw\sqrt{1-\beta^2}} = \gamma E_y \tag{13-3}$$

Electric and Magnetic Fields and Forces

The electric field perpendicular to the direction of motion is therefore increased by a factor γ.

On the other hand, if the plates were tipped up by 90° so as to be parallel to the (y,z) plane in Figure 13-1a, they would merely appear closer together in Figure 13-1b. Since the field does not depend on the plate separation, it remains unaltered in the O' frame of reference, and

$$E_x' = E_x \qquad (13\text{-}4)$$

13-2 MOVING CONDENSERS

Next consider a condenser identical to that of the last section, but moving to the right with a speed uc in the frame of reference O of Figure 13-2a. If a frame of reference O' moves to the right with speed βc, the condenser appears to have a speed $u'c$ in O'. From Equations 13-2 and 13-3 the electric and magnetic fields in the O' frame of reference are

$$E_y' = \frac{Q}{\epsilon_0 lw \sqrt{1 - u'^2}} \quad \text{and} \quad B_z' = \frac{\mu_0 Q u' c}{lw \sqrt{1 - u'^2}} \qquad (13\text{-}5)$$

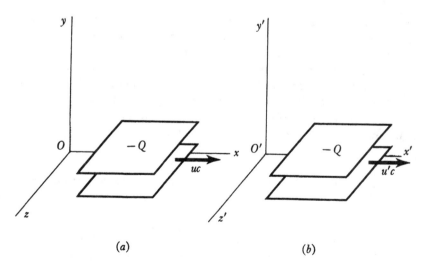

Figure 13-2

Substituting $u' = \dfrac{u - \beta}{1 - \beta u}$ from the velocity addition formula

$$E_y' = \frac{Q}{\epsilon_0 lw}\left(\frac{1-\beta u}{\sqrt{1-\beta^2}\sqrt{1-u^2}}\right)$$

$$= \gamma\left(\frac{Q}{\epsilon_0 lw\sqrt{1-u^2}} - \beta\frac{uQ}{\epsilon_0 lw\sqrt{1-u^2}}\right)$$

$$= \gamma\left(E_y - \frac{\beta B_z}{\epsilon_0 \mu_0 c}\right)$$

$$B_z' = \gamma(B_z - \beta\epsilon_0\mu_0 c E_y) \tag{13-6}$$

obtained with similar algebra. The same procedure can be followed to get E_z' and B_y'. It can also be shown that $B_x' = B_x$. It can also be demonstrated that $\epsilon_0\mu_0 = 1/c^2$. Finally, then

$$\begin{array}{ll} E_x' = E_x & cB_x' = cB_x \\ E_y' = \gamma(E_y - \beta cB_z) & cB_y' = \gamma(cB_y + \beta E_z) \\ E_z' = \gamma(E_z + \beta cB_y) & cB_z' = \gamma(cB_z - \beta E_y) \end{array} \tag{13-7}$$

Equation 13-7 describes how electric and magnetic fields transform into one another in different frames of reference. The equation bears a striking resemblance to the Lorentz transformation, but it is not identical to it. Neither the electric field nor the magnetic field is a four-vector. Rather the electric and magnetic fields make up a single quantity called the electromagnetic field tensor.

The work of these sections is by no means a proof of Equation 13-7 but only an illustration of the way in which these equations arise.

13-3 THE FIELD OF A MOVING POINT CHARGE

As a further example of the meaning of Equation 13-7, consider the field of a moving point charge. Figure 13-3a shows a point charge Q which is stationary at the origin. At a point (x,y,z) it produces an electric field given by Coulomb's Law:

Electric and Magnetic Fields and Forces

$$E_x = \frac{Q}{4\pi\epsilon_0 r^2} \cdot \frac{x}{r} = \frac{Q}{4\pi\epsilon_0} \frac{x}{(x^2 + y^2 + z^2)^{\frac{3}{2}}}$$

$$E_y = \frac{Q}{4\pi\epsilon_0} \frac{y}{(x^2 + y^2 + z^2)^{\frac{3}{2}}}$$

$$E_z = \frac{Q}{4\pi\epsilon_0} \frac{z}{(x^2 + y^2 + z^2)^{\frac{3}{2}}} \tag{13-8}$$

The magnetic field is zero since the charge is stationary.

In a frame of reference O' moving to the right with a speed βc, the charge Q appears to be moving to the left with speed βc as shown in Figure 13-3b. It is at the origin $x' = y' = z' = 0$ at time $t' = 0$. The field of this moving charge is computed by applying Equation 13-7 to 13-8. This must be done with care! We wish to know the electric and magnetic fields at a point $x' = X$, $y' = Y$, $z' = Z$ at the time $t' = 0$. This space-time point or "event" corresponds to a point (x,y,z,t) in O:

$$x = \gamma(X + \beta c \cdot 0) = \gamma X$$
$$y = Y \qquad z = Z$$
$$ct = \gamma(0 + \beta X) = \gamma\beta X \tag{13-9}$$

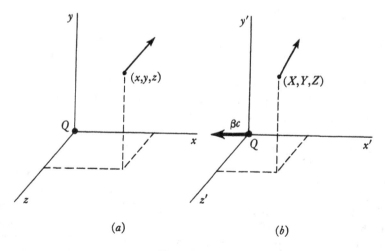

Figure 13-3

From Equations 13-8 and 13-9 the electric fields at this (x,y,z) point in O are

$$E_x = \frac{Q}{4\pi\epsilon_0} \frac{\gamma X}{(\gamma^2 X^2 + Y^2 + Z^2)^{3/2}}$$

$$E_y = \frac{Q}{4\pi\epsilon_0} \frac{Y}{(\gamma^2 X^2 + Y^2 + Z^2)^{3/2}}$$

$$E_z = \frac{Q}{4\pi\epsilon_0} \frac{Z}{(\gamma^2 X^2 + Y^2 + Z^2)^{3/2}} \quad (13\text{-}10)$$

The time ct does not enter, since the charge is stationary and these are the fields at *all* times. The electric and magnetic fields, at the corresponding (x',y',z') point in the O' frame, are now obtained by applying Equation 13-7 to 13-10.

$$E_x' = E_x = \frac{Q}{4\pi\epsilon_0} \frac{\gamma X}{(\gamma^2 X^2 + Y^2 + Z^2)^{3/2}} \qquad cB_x' = cB_x = 0$$

$$E_y' = \gamma E_y = \frac{Q}{4\pi\epsilon_0} \frac{\gamma Y}{(\gamma^2 X^2 + Y^2 + Z^2)^{3/2}}$$

$$cB_y' = \gamma\beta E_z = \frac{Q}{4\pi\epsilon_0} \frac{\gamma\beta Z}{(\gamma^2 X^2 + Y^2 + Z^2)^{3/2}}$$

$$E_z' = \gamma E_z = \frac{Q}{4\pi\epsilon_0} \frac{\gamma Z}{(\gamma^2 X^2 + Y^2 + Z^2)^{3/2}}$$

$$cB_z' = -\gamma\beta E_y = \frac{Q}{4\pi\epsilon_0} \frac{-\gamma\beta Y}{(\gamma^2 X^2 + Y^2 + Z^2)^{3/2}} \quad (13\text{-}11)$$

The electric field of the moving charge is no longer a simple inverse square field. If we write the electric field as a vector with **i**, **j**, and **k** being unit vectors in the x-, y-, and z-directions respectively

$$\mathbf{E}' = \frac{\gamma Q}{4\pi\epsilon_0(\gamma^2 X^2 + Y^2 + Z^2)^{3/2}} (\mathbf{i}X + \mathbf{j}Y + \mathbf{k}Z)$$

The vector $\mathbf{i}X + \mathbf{j}Y + \mathbf{k}Z$ is just the radius vector **r** equal in magnitude to the distance from the charge to the point (X,Y,Z) and directed radially outwards. The electric field of a moving charge is therefore still a radial field pointing directly away from the charge. It does not, however, have the same value in all directions. For example, at a point $(1,0,0)$, i.e., one meter behind the charge, the

Electric and Magnetic Fields and Forces

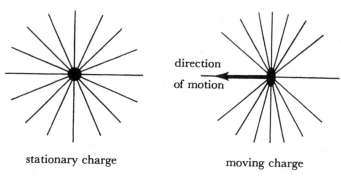

stationary charge   moving charge

Figure 13-4

field is $E_x = Q/4\pi\epsilon_0\gamma^2$. At a point (0,1,0), i.e., one meter out to the side, $E_y = \gamma Q/4\pi\epsilon_0$, or the field is γ^3 times as large as it is in front or in back of the charge. The electric field lines look like those of Figure 13-4.

The magnetic field has no component in the direction of the motion, and the magnetic field lines are concentric circles around the direction of motion. Qualitatively they behave just as in non-relativistic theory; only the magnitude is different. Notice, however, that Equation 13-11 gives the electromagnetic field of a charge moving toward *negative* x with velocity βc. It also gives the field for a single time, $t' = 0$, when the charge is at the origin. At later and earlier times, the field configuration simply follows the charge along in its motion. For example, the x-component of the electric field at a time $t' = T$ is

$$E_x' = \frac{Q}{4\pi\epsilon_0} \frac{\gamma(X + \beta c T)}{[\gamma^2(X + \beta c T)^2 + Y^2 + Z^2]^{\frac{3}{2}}} \tag{13-12}$$

13-4 GAUSS'S LAW FOR A MOVING CHARGE

In all the previous sections we have explicitly assumed that the electric charge Q remains the same in frames of reference moving with respect to one another, i.e., is a Lorentz scalar. It is not obvious that this should be the case. The other inverse r^2 field familiar to elementary physics, the gravitational field, has as its source the energy of the particle (see Section 9-7E), which is the fourth component of a four-vector. The laws by which it transforms are correspondingly different from those for the electrical fields; there is no gravitational magnetism, for example. But what is meant by the charge remaining invariant? How do we know that a certain region in space is inhabited by charges? A charge has no color; if it has a mass, it is no different from any other. The only way one can tell whether a charge exists inside a certain closed volume or not is to measure the electric field everywhere just outside the volume. If the field sticks out all over the volume, there is a charge inside; if not, there is no charge (see Figure 13-5). In other words, there is a charge Q inside a volume if the total electric flux (normal component of E times the area) is Q/ϵ_0 when added up over the surface of the volume. In the final analysis, then, Gauss's Law serves to *define* the existence of a charge.

To say that the charge remains invariant, is to say that the electric flux over a surface surrounding the charge remains invariant. In Figure 13-6a a stationary charge is shown surrounded by an imaginary cylindrical surface. The flux over the flat front and back

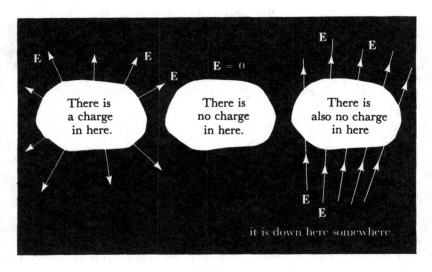

Figure 13-5

Electric and Magnetic Fields and Forces

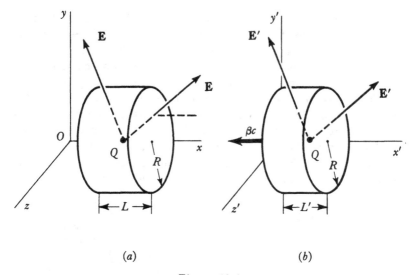

Figure 13-6

surfaces we call Φ_f and over the curved outer surface Φ_c. Gauss's Law states that $\Phi_f + \Phi_c = Q$ *no matter what the shape of the cylinder is*. It may be very long and skinny with a large curved surface and almost no ends, or short and fat. In other words, the ratio Φ_f/Φ_c is quite arbitrary; only the sum is fixed.

In a frame of reference O', where the charge is moving, the imaginary cylinder appears contracted, as in Figure 13-6b. If, in this new frame of reference, the charge remains Q, then $\Phi_f' + \Phi_c' = Q = \Phi_f + \Phi_c$. Furthermore, since the ratio Φ_f/Φ_c is *arbitrary*, the only way this can happen is that $\Phi_f' = \Phi_f$ and $\Phi_c' = \Phi_c$. Consider first the front and back. Φ_f involves *only* the product of the area and the *x-component* of the electric field. Since the front and back disks remain the same size in O and O', and since $\Phi_f = \Phi_f'$, the x-component of the field must remain the same: $E_x' = E_x$. Now consider the flux over the curved side, which involves *only* the component of the field perpendicular to the x-axis. It must remain the same, i.e., $\Phi_c = \Phi_c'$. However, the length of the cylinder and hence the area of the curved side *shrinks* by the factor γ. Therefore the field components perpendicular to the x-axis must *increase* by the factor γ, or $E_y' = \gamma E_y$ and $E_z' = \gamma E_z$. Therefore, at least in the case where the charge is stationary, and $B = 0$, Equation 13-7 is a consequence of the fact that charge is an invariant.

The argument that charge invariance requires a magnetic field is a bit more involved. Suppose the charge is at rest in the O system, as before, and there are, therefore, electric but no magnetic fields. The O' system moves, as before, with velocity βc along the x-axis. We also introduce a third system O'' moving at a velocity uc along the x'-axis with respect to O' and, therefore, $vc = c(u + \beta)/(1 + \beta u)$ with respect to O. From the preceding invariance argument, which works for the O'' system as well as the O' system, we have

$$E_y' = \frac{E_y}{\sqrt{1-\beta^2}} \qquad E_y'' = \frac{E_y}{\sqrt{1-v^2}} = \frac{E_y}{\sqrt{1-\left(\frac{u+\beta}{1+\beta u}\right)^2}}$$

$$= \frac{(1+\beta u)E_y}{\sqrt{1-u^2}\sqrt{1-\beta^2}} \quad (13\text{-}13)$$

We now attempt to write E_y'' as a transformation from the O' system, i.e., in terms of quantities in the O' system and the velocity of the O'' system with respect to the O' system. It must *not* involve E_y or β. From Equation 13-13 this looks impossible. We can eliminate E_y or β, but not both. However, by making use of the magnetic field in O', it can be done. Equation 13-13 becomes

$$E_y'' = \frac{E_y}{\sqrt{1-u^2}\sqrt{1-\beta^2}} + \frac{\beta u E_y}{\sqrt{1-u^2}\sqrt{1-\beta^2}}$$

$$= \frac{1}{\sqrt{1-u^2}}\left(E_y' + u\frac{\beta E_y}{\sqrt{1-\beta^2}}\right) \quad (13\text{-}14)$$

Since, from Equation 13-2, $cB_z = -\gamma \epsilon_0 \mu_0 \beta c^2 E_y = -\gamma \beta E_y$

$$E_y'' = \frac{1}{\sqrt{1-u^2}}(E_y' - ucB_z')$$

This is identical to the second part of Equation 13-7.

Although we recognized the magnetic field term in Equation 13-14 and inserted it, it *could* have been introduced as a theoretical necessity for making Coulomb's Law a relativistically acceptable

Electric and Magnetic Fields and Forces

law and the charge an invariant quantity. Historically, of course, things were quite the other way about. Magnetic forces were known long before it was even suspected they were connected to electrical phenomena. Next they were identified as a result of the motion of charge. Finally, because the form of Equation 13-7 was known, it was possible to identify charge as a Lorentz invariant! In fact, it is very difficult to imagine how the theory of relativity itself could have been discovered, had not electricity and magnetism grown up previously as a highly developed experimental and theoretical science.

13-5 ELECTRICAL FORCES

Classically we know what will happen to a charged particle moving slowly through electric and magnetic fields. When the motion becomes rapid, the answer is not so obvious. However, there is a scheme for finding out. Suppose such a rapidly moving charge q moves with a speed βc through some combination of electric and magnetic fields $\mathbf{E}$ and $\mathbf{B}$. Without loss of generality, the x-axis may be chosen to be the direction of the particle's motion. We may then choose to observe this motion in a frame of reference O' moving along the x-axis with velocity βc. That is, in O' the particle is momentarily at rest. We are now faced with a classical problem which we know how to solve. In a very short time, from $t' = 0$ to $t' = \Delta t'$, the particle accelerates from a momentum $p = 0$ to a momentum $p_x' = \Delta p_x' = qE_x'\Delta t'$ and $p_y' = \Delta p_y' = qE_y'\Delta t'$. The z-direction has been momentarily dropped for convenience, and we have somewhat anticipated our final result by giving the classical results in terms of momentum, though these changes of momentum in O' could just as well have been written as mass times acceleration. The energy of the particle at rest is $e' = mc^2$, and after the acceleration it is still mc^2.[1]

---

[1] Actually, of course, it has some kinetic energy. It may be included with no harm, but considerable additional work. In the end it turns out to be of second order and drops out without changing the argument. The same thing can be said for the neglect of magnetic forces in this frame of reference. They are zero when the particle is at rest, but should play some role as it gets moving. However, if $\Delta t'$ remains arbitrarily small the effect of the magnetic forces can be made negligible.

We now calculate what this process looks like back in the original frame of reference. From a Lorentz transformation we find that at $t = 0$ the particle had momentum

$$cp_x = \gamma(cp_x' + \beta e') = \gamma\beta mc^2$$

which is, of course, just the momentum of a particle of mass m moving with velocity βc. Also

$$cp_y = cp_y' = 0$$

After the acceleration at a time $t = \Delta t$

$$cp_x = \gamma(cqE_x'\Delta t' + \beta mc^2)$$
$$cp_y = cqE_y'\Delta t'$$

The change in momentum during the acceleration was therefore

$$\Delta p_x = \gamma q E_x' \Delta t'$$
$$\Delta p_y = q E_y' \Delta t'$$

The particle was nearly at rest in the primed system, and therefore $\Delta t'$ is a proper time and $\Delta t = \gamma \Delta t'$. The rate of change of momentum with respect to time $\Delta p / \Delta t$ is therefore

$$\frac{\Delta p_x}{\Delta t} = qE_x' \qquad \frac{\Delta p_y}{\Delta t} = \frac{1}{\gamma} qE_y' \qquad (13\text{-}15)$$

Using Equation 13-7 to express E' in terms of the fields in the original frame of reference, Equation 13-15 becomes

$$\frac{\Delta p_x}{\Delta t} = qE_x$$
$$\frac{\Delta p_y}{\Delta t} = q(E_y - \beta c B_z)$$
$$\frac{\Delta p_z}{\Delta t} = q(E_z + \beta c B_y) \qquad (13\text{-}16)$$

where the z-equation can be similarly derived. If the relativistic force is defined to be this time rate of change of momentum, Equation

Electric and Magnetic Fields and Forces

13-16 becomes, in vector form,

$$\mathbf{F} = \frac{d\mathbf{p}}{dt} = q(\mathbf{E} + \mathbf{v} \times c\mathbf{B}) \tag{13-17}$$

This is just the familiar classical force expression for the motion of a charged particle in electric and magnetic fields.

13-6 RELATIVISTIC FORCE

"A particle at rest will remain at rest, and a particle in motion will remain in uniform rectilinear motion unless acted on by a force." So runs a familiar statement of Newton's or Galileo's First Law. "Rest" and "uniform motion" are relativistically exactly the same state, merely observed from different reference frames. A *force* makes its presence known by a change in this state. Classically, the force is usually taken to be (mass) × (acceleration), the acceleration describing the change from uniform motion. Since mass is necessarily constant in classical physics: $F = ma = m(dv/dt) = d(mv)/dt = dp/dt$. Mass times acceleration, and rate of change of momentum are, therefore, equally good as classical definitions of force.

Because of the proper-improper time difference in special relativity there are several possible ways of defining force. If t is the time in the frame of reference in which the motion is being described, and τ is the time measured by clocks on the particle, i.e., the proper time, then these definitions are possible:

1. $F_x = m \dfrac{d^2 x}{dt^2}$

2. $F_x = m \dfrac{d}{dt}\dfrac{dx}{d\tau} = \dfrac{dp_x}{dt}$

3. $F_x = m \dfrac{d^2 x}{d\tau^2} = \dfrac{dp_x}{d\tau}$

All three describe a change away from uniform motion with time. The first has been used in the past because of its close relation to the classical expression. It leads to awkward expressions and is almost never used today. The third is in most ways the most natural expression. It is called the Minkowski force. It is a four-vector as can be seen from the fact that **p** and therefore **dp** is a four-vector, whereas $d\tau$ is a proper time, a four-scalar. A vector divided by a scalar is a vector. Because of this property, it is most useful in theoretical discussions.

The second definition mixes proper and improper time intervals in its definition and therefore seems the most unlikely candidate of all. Nevertheless it is the most popular choice, for which there are several reasons. One is that it is a good pedagogical choice, since Newtonian mechanics can be written in the same form, and the step to relativistic mechanics is thereby made easier. Many problems can be worked in the familiar classical way by simply substituting $p = mv/\sqrt{1 - v^2/c^2}$ for the classical momentum. Too much is often made of this. Mass simply must be conserved if classical physics is to be a self-consistent theory. Probably the most compelling reason to choose the relativistic force to be dp/dt is Equation 13-17; the familiar form of the classical electromagnetic force law is preserved. No matter how one talks about the general concept of force in relativity, it is a fact that the only practical applications (known to the author) are forces in electric and magnetic fields. Gravitational forces are properly treated only in general relativity, and nuclear forces are almost never treated except in quantum theory, where the concept of force plays a minor role. For such practical applications dp/dt seems the most convenient choice. It should be realized, however, that it is not *a priori* obvious what choice to make. There are several possible choices; several have actually been used, and the choice made is dictated as much by convenience as anything else.

13-7 CENTRIPETAL FORCE AND MAGNETIC DEFLECTION

Suppose a relativistic particle travels in a circle with uniform speed vc and therefore uniform magnitude of momentum p. What force must act on it? In Figure 13-7a the particle is shown at two neighboring points on its circular path of radius r, which it reaches at times Δt apart. The distance $\overline{MN} = r\Delta\theta$, so that $r\Delta\theta = vc\Delta t$.

Electric and Magnetic Fields and Forces

During this time, its momentum remains at constant magnitude p, but is also deflected through the angle $\Delta\theta$. As shown in Figure 13-7b, the change in momentum $\Delta p = p\Delta\theta$ and is perpendicular to the momentum. It follows that $r\Delta p = pvc\Delta t$ or

$$\frac{\Delta p}{\Delta t} = \frac{pcv}{r} = \begin{pmatrix}\text{centripetal}\\\text{force}\end{pmatrix} \qquad (13\text{-}18)$$

Since $\Delta p/\Delta t$ is the force acting on the particle, we conclude that the centripetal force necessary to keep a particle moving in a circle is pcv/r. Everything done up to this point is identical in the relativistic and classical treatments so that Equation 13-18 also gives the familiar classical expression if mvc is substituted for p.

Probably the most useful application of this centripetal force formula is in calculating the radius of curvature of a charged particle in a magnetic field. Suppose such a particle of charge q enters a magnetic field $\mathbf{B}$ with a velocity vc and momentum $\mathbf{p}$ perpendicular to $\mathbf{B}$. The magnetic force $\mathbf{F} = q\mathbf{v} \times c\mathbf{B}$ is always perpendicular to $\mathbf{v}$ and hence to $\mathbf{p}$. The momentum therefore never changes its magnitude, but is merely deflected. The force, therefore, never changes its magnitude, but remains $qvcB$ perpendicular to the momentum and $\mathbf{B}$. This magnetic force, therefore, constitutes the centripetal force necessary to swing the particle in a circle where

$$F = qvcB = pcv/r$$

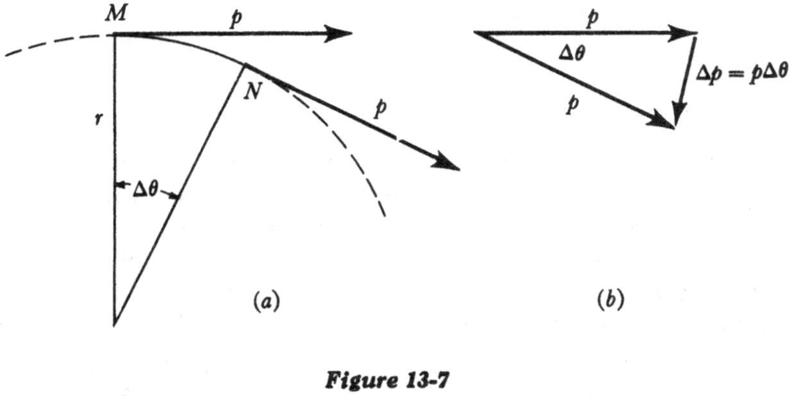

Figure 13-7

Table 13-1. *Ratios of Experimental to Theoretical Momenta*

Measured v/c	Measured momentum ÷ classically predicted momentum	Measured momentum ÷ relativistically predicted momentum
.3173	1.057	.995
.3787	1.074	1.000
.4281	1.097	.999
.5154	1.142	1.001
.6870	1.271	1.003

or

$$r = \frac{pc}{qcB} \tag{13-19}$$

Equation 13-19 has a certain historical significance, because it was by observing the deflection of high speed electrons from radioactive disintegrations in a magnetic field that the first mechanical consequences of the theory of relativity were observed.[2] Electrons of a given speed were prepared by passing them through a velocity selector (see Exercise 5) and then into a uniform magnetic field where the radius of curvature of their path was measured. In terms of the particles' speed, Equation 13-19 becomes

$$r = \frac{mvc}{qcB\sqrt{1-v^2}} \text{ (relativistically)}$$

$$= \frac{mvc}{qcB} \text{ (classically)}$$

At the time that the measurements were made, it was somewhat uncertain that β-rays were electrons, and the experimental results were quoted as measurements of q/m of the radioactive particles.

Furthermore, Lorentz had postulated that the mass of an electron was entirely electromagnetic in nature, and from electromagnetic theory alone had predicted the relativistic form of the momentum.

[2] A. H. Bucherer, *Ann. Physik*, **28**, 513 (1909). See also J. D. Stranathan, *The "Particles" of Modern Physics*, Blakiston, Philadelphia (1942), p. 138.

Electric and Magnetic Fields and Forces

The results were therefore quoted as proof of the electromagnetic nature of the mass of the electron, rather than as confirmation of Einstein's postulates. However, if we recast the results as measuring the momentum of the electrons, they are given in Table 13-1. The results speak for themselves.

Exercises

1. Transform the E' and B' fields given in Equation 13-7 back to the original system by applying Equation 13-7 *again*, but with β replaced by $-\beta$.
2. Solve Equation 13-7 for E and B in terms of E' and B'. How do you know your answer is correct?
3. Show that $E^2 - c^2B^2$ is an invariant.
4. Consider a frame of reference where there is a condenser moving with velocity uc to the *right* as in Section 13-2. In this same frame of reference there is an identical condenser oppositely charged moving *left* with velocity uc. Consider that the "plates" penetrate each other. This imaginary setup is to make a model of a situation in which there is no net electric charge, but there is a current, and therefore a magnetic field between the plates.
   a. Find this magnetic field between the plates.
   b. In a frame of reference where the whole device is moving left with speed βc, find the electric and magnetic fields by using Equation 13-7.
   c. Where does the electric field come from? That is, explain how the initially electrically neutral system appears to be charged in a moving frame of reference. Does this violate the concept that charge is an invariant?
5. Suppose a region of space has an electric field in the y-direction $E_y = E$, and a magnetic field $B_z = B$ perpendicular to it. Such a device is often made by inserting a long condenser between the poles of a magnet. If a particle of charge q moves between the condenser plates with a particular velocity in the x-direction, the magnetic and electric forces cancel. Only particles with this particular velocity feel no force and are undeflected. Such a device can therefore serve as a velocity selector for charged particles.
   a. In terms of E and B, what velocity particle will be undeflected?
   b. Look at this device from a frame of reference moving with the velocity βc in the x-direction. From Equation 13-7 what new E and B fields exist in the device?
   c. What velocity particle in this new frame of reference will be undeflected?
   d. Discuss how you know your answer is right.
6. Suppose a particle of charge q is released from rest in the perpendicular fields $E_y = E$ and $B_z = B$ of Exercise 5. At first sight the motion seems

complicated. The particle will start to move in the direction of the electric field, but as soon as it attains a velocity it will be deflected by the magnetic field in a constantly changing radius of curvature. However, there is a frame of reference where the motion is very simple, namely the frame where the electric field is zero. Give all answers in terms of E, B, q, and m.

   a. With what velocity is this frame of reference moving?

   b. With what velocity is the particle moving in this frame of reference? With what momentum?

   c. What is the magnetic field in this frame of reference?

   d. Since there is no electric field, but only a constant magnetic field in this frame of reference, the path of the particle is a circle. What is its radius?

   e. What will be the path of the particle in the original frame of reference? First ignore relativistic effects, and then consider what difference they make.

   f. Calculate the momentum and energy of the particle in the moving frame when it has got to the other side of the circle from which it started. At this point what will be its energy in the original frame of reference? What has been its *gain* in energy?

   g. Compare the gain in energy calculated in (*f*) with the work done by the field, i.e., the force in the y-direction times the displacement in the y-direction.

Appendix A

Approximate Calculations in Relativity

IN PHYSICS, AND ESPECIALLY IN RELATIVISTIC PHYSICS, it is often necessary to evaluate expressions like $x = (y + z)^n$ where $y \gg z$. If y is really so very much larger than z that z can be neglected entirely, then, of course, x is approximately y^n. Very frequently, however, the expression occurs in combination with another, like $(y + z)^n - y^n$. If z is neglected entirely, the expression is zero. This may be good enough; it may not. If not, the following technique is useful:

1. Multiply by y^n and divide by y^n.

$$x = (y + z)^n = y^n \frac{(y + z)^n}{y^n} = y^n \left[1 + \left(\frac{z}{y}\right)\right]^n$$

2. Expand the expression in brackets by the binomial theorem:

$$x = y^n \left[1 + \left(\frac{z}{y}\right)\right]^n = y^n \left[1 + n\left(\frac{z}{y}\right) + \frac{n(n-1)}{2}\left(\frac{z}{y}\right)^2 + \cdots\right]$$

3. Notice that since $z \ll y$, $(z/y) \ll 1$, and the successive terms in the brackets get smaller and smaller very rapidly. Neglect all

those too small to affect the answer you want. Neglecting all but the first gives $x = y^n$ as before. If more accuracy is necessary, keep the second; thus, for the example given above:

$$(y + z)^n - y^n \approx y^n \left[1 + n\left(\frac{z}{y}\right)\right] - y^n \approx ny^n\left(\frac{z}{y}\right)$$

A frequent example in relativity is $(1 - v^2/c^2)^{-\frac{1}{2}}$. Thus

$$\left(1 - \frac{v^2}{c^2}\right)^{-\frac{1}{2}} \approx 1 - \frac{1}{2}\left(-\frac{v^2}{c^2}\right) \approx 1 + \frac{1}{2}\left(\frac{v^2}{c^2}\right)$$

Inexperienced readers should probably try a few examples numerically. Thus

$$\sqrt{1.1} = 1.04881 \ldots$$

whereas the approximate formula keeping two terms gives

$$\sqrt{1.1} \approx 1 + \tfrac{1}{2}(.1) = 1.05$$

Another example which arises frequently is the slide rule evaluation of the difference between the squares of two nearly equal numbers. Thus, to calculate the momentum of a pion (mc^2 = 139.6 MeV) of energy $E = 141.2$ MeV we must evaluate

$$\begin{aligned} pc &= \sqrt{E^2 - m^2c^4} \\ &= \sqrt{141.2^2 - 139.6^2} \\ &= \sqrt{19937.44 - 19488.16} = \sqrt{449.28} \end{aligned}$$

If the squaring were done on the slide rule, the answer would likely be quite inaccurate. On the other hand the expression can be factored. Thus

$$\begin{aligned} pc &= \sqrt{E^2 - m^2c^4} = \sqrt{(E + mc^2)(E - mc^2)} \\ &= \sqrt{(141.2 + 139.6)(141.2 - 139.6)} \\ &= \sqrt{280.8 \times 1.6} = \sqrt{449.28} \end{aligned}$$

Not only is slide rule accuracy preserved, but the manipulation is so much simpler that this is a useful technique even when the two numbers differ considerably in size.

Appendix B

A Summary of Relativistic Formulas

IN THE FOLLOWING FORMULAS all c's have been deleted, i.e., all times are given in units of distance, velocities as fractions of the velocity of light, all momenta, masses, and energies in energy units. To "restore the c's":

1. Multiply each time by c.
2. Divide each velocity by c.
3. Multiply each momentum by c.
4. Multiply each mass by c^2.

The velocity u refers to the motion of an object in a particular frame of reference. The velocity β and $\gamma = (1 - \beta^2)^{-\frac{1}{2}}$ refers to the relative velocity between coordinate systems—although it must be remembered that this artificial distinction cannot always be maintained. For example, if an object moves through the system O with velocity u and one wishes to choose a second system O' in which the object is at rest, then the relative velocity between systems is $\beta = u$.

The primed system moves along the positive x-axis of the unprimed system. The z-direction always behaves like the y-direction and is not given explicitly in these formulas.

I. Basic Time and Length Measurements

    a. (Proper time) = (Improper time) $\sqrt{1 - \beta^2}$.

    b. If an object of length L moves with velocity β parallel to its length through some frame of reference, a simultaneous measurement of its length in this frame of reference will give

$$L' = L\sqrt{1 - \beta^2}$$

    c. If an object of length L moves in a direction perpendicular to its length through some frame of reference, a measurement of its length in this frame of reference will give

$$L' = L$$

II. Velocity Addition Formulas

$$u_x' = \frac{u_x - \beta}{1 - \beta u_x} \qquad u_y' = \frac{u_y \sqrt{1 - \beta^2}}{1 - \beta u_x}$$

III. Lorentz Transformations

$$x' = \gamma(x - \beta t)$$
$$y' = y$$
$$t' = \gamma(t - \beta x)$$

where

$$\gamma = \frac{1}{\sqrt{1 - \beta^2}} \qquad \beta^2 = 1 - \frac{1}{\gamma^2} \qquad \gamma^2 - \beta^2 \gamma^2 = 1$$

IV. Proper Velocity

$$\eta_x = \frac{u_x}{\sqrt{1 - u^2}} \qquad \eta_y = \frac{u_y}{\sqrt{1 - u^2}} \qquad \eta_0 = \frac{1}{\sqrt{1 - u^2}}$$

$$\eta_x' = \gamma(\eta_x - \beta \eta_0)$$
$$\eta_y' = \eta_y$$
$$\eta_0' = \gamma(\eta_0 - \beta \eta_x)$$

A Summary of Relativistic Formulas

V. Momentum and Energy

$$p_x = \frac{mu_x}{\sqrt{1-u^2}} \qquad p_y = \frac{mu_y}{\sqrt{1-u^2}} \qquad E = \frac{m}{\sqrt{1-u^2}}$$

$$p = uE \qquad E^2 - p^2 = m^2$$
$$p_x' = \gamma(p_x - \beta E)$$
$$p_y' = p_y$$
$$E' = \gamma(E - \beta p_x)$$
$$E'^2 - p'^2 = E^2 - p^2 = m^2$$

VI. Four-Vectors

$$A_x' = \gamma(A_x - \beta A_0)$$
$$A_y' = A_y$$
$$A_0' = \gamma(A_0 - \beta A_x)$$
$$A'^2 = A_0'^2 - A_x'^2 - A_y'^2 = A_0^2 - A_x^2 - A_y^2 = A^2$$
$$\mathbf{A'B'} = A_0'B_0' - A_x'B_x' - A_y'B_y' = A_0B_0 - A_xB_x - A_yB_y = \mathbf{AB}$$

Appendix C

A Table of Particles

THIS TABLE IS INCLUDED for convenience in working problems. It is not a complete table of all particles or resonant states presently known and, even for those particles shown, only the principal decay modes are given.

In the case of charged particles, only one sign of the charge is given. There is always an antiparticle of opposite sign whose decay products are of opposite charge sign. In the case of the γ^0, π^0, ρ^0, and ω^0, the antiparticle is the *same* as the particle. In the case of the ν^0, n^0, Λ^0, Σ^0, the antiparticle is also uncharged, but is distinct. For example, a neutron decays into a proton, e^-, and antineutrino. An antineutron cannot do this, although charge would still be conserved; it must decay into an antiproton, e^+, and neutrino.

When particles are produced in collisions of other particles, neutrinos, electrons, muons, protons, neutrons, lambdas, and sigmas must be produced in pairs, i.e., the particle and its antiparticle. The others may be produced singly.

Finally, the question of whether neutral kaons are their own antiparticles or not is too complicated to be outlined here, but is probably understood in its important aspects. Apparently there are two kinds of neutrinos, one which goes with muons and one which goes with electrons. The significance of this is not presently understood.

A Table of Particles

Table of Particles

Particle	Symbol	Mass (MeV)	Decay products	Mean life (seconds)
photon	γ^0	0	stable	–
neutrino	ν^0	0	stable	–
electron	e^{-}	.511	stable	–
muon	μ^-	105.6	$(e^-\nu\nu)$	2.21×10^{-6}
pion	π^+	139.6	$(\mu^+\nu)$	2.55×10^{-8}
	π^0	135.0	$(\gamma\gamma)$	1×10^{-16}
kaon	K^+	493.9	$(\mu^+\nu)(\pi^+\pi^0)(\mu^+\pi^0\nu)$ $(e^+\pi^0\nu)(\pi^+\pi^+\pi^-)(\pi^+\pi^0\pi^0)$	1.22×10^{-8}
	K_1^0	497.8	$(\pi^+\pi^-)(\pi^0\pi^0)$	1.0×10^{-10}
	K_2^0	497.8	$(\pi^\pm e^\mp \nu)(\pi^\pm \mu^\mp \nu)(\pi^+\pi^-\pi^0)(\pi^0\pi^0\pi^0)$	$\sim 6 \times 10^{-8}$
eta meson	η^0	548	$(\pi^+\pi^-\pi^0)(\pi^0\pi^0\pi^0)(\pi^+\pi^-\gamma)(\gamma\gamma)$	short
rho meson	ρ^+	763	$(\pi^+\pi^0)$	short
	ρ^0		$(\pi^+\pi^-)$	
omega meson	ω^0	782	$(\pi^+\pi^-\pi^0)(\pi^0\gamma)$	short
proton	p^+	938.2	stable	–
neutron	n^0	939.5	$(p^+e^-\nu)$	1.01×10^3
lambda	Λ^0	1115.4	$(p\pi^-)(n\pi^0)$	2.5×10^{-10}
sigma	Σ^+	1189	$(n\pi^+)(p\pi^0)$	$.8 \times 10^{-10}$
	Σ^0	1193	$(\Lambda\gamma)$	short
	Σ^-	1197.4	$(n\pi^-)$	1.6×10^{-10}
cascade	Ξ^0	1315	$(\Lambda^0\pi^0)$	2.8×10^{-10}
	Ξ^-	1321.2	$(\Lambda^0\pi^-)$	1.75×10^{-10}

Index*

Aberration of light, 81–83, *90*, *92*, *157*
   classical, 27–29
Absolute motion, 29
   (*see also* Velocity, absolute)
Acceleration, 8, *78*, 94
   parallel to the direction of motion, 83–86, 108–109
   perpendicular to the direction of motion, *112*
Addition of velocities (*see* Velocity, addition of)
Antiparticles, 212
Antiprotons, 169–171
Approximate calculations, 207–208

Beta (β), 105
Binary star experiment, 40, 43–45, 47

Causality, *111–112*
Center-of-mass, 159–160, 161
   (*see also* Collision)
Center-of-momentum (*see* Center-of-mass)
Centripetal force, 202–203
Charge, 195–196
   invariance of, 195–196
   moving, 192–195
Clocks, moving, 54–56
   synchronization of, 53
Collision, elastic, equal mass particles, 144–145, *149*, 164–169, 185–188
   glancing, 127–129
   head on, 145–147
   center-of mass, 164–171, 186–187
   (*see also* Explosion)
Conservation of energy (*see* Energy, conservation of)

---

* Where exercises have particular relevance to the development of a subject they have been referred to in this index. Such references are italicized.

Conservation of momentum (*see* Momentum, conservation of)
Coordinate system, 5
Coordinate transformation, 103–104, *109*, 178–180
   classical, 5–7

De Sitter experiment, 40, 43–45, *47*
Doppler effect, 88–90, *91*, 99
   classical, 22–26
Double star experiment, 40, 43–45, *47*

Electric field, 184, 189
   of moving charge, 192–195
   transformation of, 190–192
Electric force, 199–201
Energy, 132–139
   center-of-mass, 163–164
   conservation of, 132–139, *148–149*
      classical, *18*
   equivalence to mass, 134–135, 138–139
      gravitational mass experiment, 143
      inertial mass experiment, 140–143
   four-vector, 177–178
   function of speed, experiment, 140
   total, of a group of particles, 160
   transformation of, 148
   units of, 135
Eötvös experiment, 143
Equivalence of mass and energy (*see* Energy)
Ether, 27
   drag, *91*

Events, 51
   spacelike, *111–112*
   timelike, *111–112*
Explosion, example of a relativistic, 135–139

First postulate of relativity, 2, 4, 13, 41
Force, 199–202
   centripetal, 202–203
   electric, 199–201
   magnetic, 200–201, 203–205
Formulas, relativistic, 209–211
Four-scalar, 120, 182–183
Four-vector, 115, 177–178, 183–184
Four-velocity (*see* Velocity, proper)
Frame of reference
   center-of-mass, 161
   definition of, 3, 5
   inertial, 8, 100
Frequency, 21–22
Fresnel drag, *91*

Gamma (γ), 105
Gauss's Law, 195–199
Gedanken experiment, 31–32, 49–50
General relativity, 9
Graviton, 153

Inertia, law of, 8
Inertial frames of reference (*see* Frame of reference)

K-meson decay, 141–143, 155–157, *157*, *173*

Index

Length, paradox, 68–72
  parallel to motion, contraction of, 63–64, 66–68, 107–108
  perpendicular to motion, 73–74
Light, aberration of (*see* Aberration)
  constancy of the speed of, 2, 46, *91*
  (*see also* Postulates of relativity, second)
  emission of, by moving objects (*see* Aberration)
  speed of, 26–27, 87
  waves, 26–27
Lorentz-Fitzgerald contraction, 40
  (*see also* Length)
Lorentz invariant, 120, 182–183
Lorentz scalar, 120, 182–183
Lorentz transformation, 103–104, *109*, 178–180

Magnetic field, 189, 198–199
  of a moving charge, 192–195
  transformation of, 190–192
Magnetic force, 200–201, 203–205
Magnetic monopole, *171*
Mass, conservation of, classical, 126
  electromagnetic, 204
  equivalence to energy (*see* Energy, equivalence to mass)
Massless particles, 151–152
Medium, for propagation of waves, 19–20
MeV, 135
Michelson-Morley experiment, 34–41, *46*, *47*, 64–66
Minkowski force, 202
Momentum, 126–130, 204–205
  conservation of, 130–132, 137–138
  classical, 9–12, *15–18*, 124–126

Momentum, four-vector, 177–178
  total, of a group of particles, 160
  transformation of, 148
Muons, time dilation in the decay of, 57–60

Neutrino, 57, 153, *157*
Newton's laws (*see* Inertia, law of)

Omega minus (Ω^-) particle, *173–176*

Particles, table of some properties of elementary, 212–213
Periodic waves, 21–22
Photons, 152–153
Pions,
  decay, 155–157, *157*
  speed of light from the decay of neutral, 43, 45–46
Planck's constant, 153
Positrons, annihilation of, 43, 45
Postulates of relativity,
  first, or relativity postulate, 2, 4, 13, 41
  second, or speed of light postulate, 2, 41–42, 46
  experimental proof of, 42–46
Proper time, 50–54, *111*, 120
Proper velocity (*see* Velocity, proper)

Quantum, 153
Quantum hypothesis, 153

Relativity, general, 9
Relativity, postulates of (*see* Postulates of relativity)
Restoring the "c's," 209
Ritz, electromagnetic theory, 40

Scalars, 180–182
   (*see also* Four-scalar)
Scattering (*see* Collision)
Second postulate of relativity (*see* Postulates of relativity)
Simultaneity, 68–72
Space-time, 177–178, 180
Speed, ultimate, 87–88
Synchronization of clocks, 53
Systems of particles, 159–160

Time, dilation, 49–60, 106–107
    experiment, 57–60
    improper, 50–54
    interval, 51
    proper, 50–54, *111*, 120
Twin paradox, 93–102
    experiment, 101–102

Units,
   consistent, for space-time and momentum-energy, 105–106, 145–146, 209
   of energy, 135

Velocity, absolute, 12–13
   (*see also* Absolute motion)
   addition of, 77–81, *109*
   classical, 12, *15*
   proper, 113–115
     addition of, or transformation of, 115–121
   relative, 12–13, 13–15
   (*see also* Velocity, addition of)
Velocity selector, electromagnetic, *205*

Waves, 19–20
   periodic, 21–22
Wavelength, 21–22

Zero mass particles, 151–152

A CATALOG OF SELECTED
DOVER BOOKS
IN SCIENCE AND MATHEMATICS

CATALOG OF DOVER BOOKS

Mathematics-Bestsellers

HANDBOOK OF MATHEMATICAL FUNCTIONS: with Formulas, Graphs, and Mathematical Tables, Edited by Milton Abramowitz and Irene A. Stegun. A classic resource for working with special functions, standard trig, and exponential logarithmic definitions and extensions, it features 29 sets of tables, some to as high as 20 places. 1046pp. 8 x 10 1/2. 0-486-61272-4

ABSTRACT AND CONCRETE CATEGORIES: The Joy of Cats, Jiri Adamek, Horst Herrlich, and George E. Strecker. This up-to-date introductory treatment employs category theory to explore the theory of structures. Its unique approach stresses concrete categories and presents a systematic view of factorization structures. Numerous examples. 1990 edition, updated 2004. 528pp. 6 1/8 x 9 1/4. 0-486-46934-4

MATHEMATICS: Its Content, Methods and Meaning, A. D. Aleksandrov, A. N. Kolmogorov, and M. A. Lavrent'ev. Major survey offers comprehensive, coherent discussions of analytic geometry, algebra, differential equations, calculus of variations, functions of a complex variable, prime numbers, linear and non-Euclidean geometry, topology, functional analysis, more. 1963 edition. 1120pp. 5 3/8 x 8 1/2. 0-486-40916-3

INTRODUCTION TO VECTORS AND TENSORS: Second Edition--Two Volumes Bound as One, Ray M. Bowen and C.-C. Wang. Convenient single-volume compilation of two texts offers both introduction and in-depth survey. Geared toward engineering and science students rather than mathematicians, it focuses on physics and engineering applications. 1976 edition. 560pp. 6 1/2 x 9 1/4. 0-486-46914-X

AN INTRODUCTION TO ORTHOGONAL POLYNOMIALS, Theodore S. Chihara. Concise introduction covers general elementary theory, including the representation theorem and distribution functions, continued fractions and chain sequences, the recurrence formula, special functions, and some specific systems. 1978 edition. 272pp. 5 3/8 x 8 1/2.
0-486-47929-3

ADVANCED MATHEMATICS FOR ENGINEERS AND SCIENTISTS, Paul DuChateau. This primary text and supplemental reference focuses on linear algebra, calculus, and ordinary differential equations. Additional topics include partial differential equations and approximation methods. Includes solved problems. 1992 edition. 400pp. 7 1/2 x 9 1/4. 0-486-47930-7

PARTIAL DIFFERENTIAL EQUATIONS FOR SCIENTISTS AND ENGINEERS, Stanley J. Farlow. Practical text shows how to formulate and solve partial differential equations. Coverage of diffusion-type problems, hyperbolic-type problems, elliptic-type problems, numerical and approximate methods. Solution guide available upon request. 1982 edition. 414pp. 6 1/8 x 9 1/4. 0-486-67620-X

VARIATIONAL PRINCIPLES AND FREE-BOUNDARY PROBLEMS, Avner Friedman. Advanced graduate-level text examines variational methods in partial differential equations and illustrates their applications to free-boundary problems. Features detailed statements of standard theory of elliptic and parabolic operators. 1982 edition. 720pp. 6 1/8 x 9 1/4. 0-486-47853-X

LINEAR ANALYSIS AND REPRESENTATION THEORY, Steven A. Gaal. Unified treatment covers topics from the theory of operators and operator algebras on Hilbert spaces; integration and representation theory for topological groups; and the theory of Lie algebras, Lie groups, and transform groups. 1973 edition. 704pp. 6 1/8 x 9 1/4.
0-486-47851-3

Browse over 9,000 books at www.doverpublications.com

CATALOG OF DOVER BOOKS

A SURVEY OF INDUSTRIAL MATHEMATICS, Charles R. MacCluer. Students learn how to solve problems they'll encounter in their professional lives with this concise single-volume treatment. It employs MATLAB and other strategies to explore typical industrial problems. 2000 edition. 384pp. 5 3/8 x 8 1/2. 0-486-47702-9

NUMBER SYSTEMS AND THE FOUNDATIONS OF ANALYSIS, Elliott Mendelson. Geared toward undergraduate and beginning graduate students, this study explores natural numbers, integers, rational numbers, real numbers, and complex numbers. Numerous exercises and appendixes supplement the text. 1973 edition. 368pp. 5 3/8 x 8 1/2. 0-486-45792-3

A FIRST LOOK AT NUMERICAL FUNCTIONAL ANALYSIS, W. W. Sawyer. Text by renowned educator shows how problems in numerical analysis lead to concepts of functional analysis. Topics include Banach and Hilbert spaces, contraction mappings, convergence, differentiation and integration, and Euclidean space. 1978 edition. 208pp. 5 3/8 x 8 1/2. 0-486-47882-3

FRACTALS, CHAOS, POWER LAWS: Minutes from an Infinite Paradise, Manfred Schroeder. A fascinating exploration of the connections between chaos theory, physics, biology, and mathematics, this book abounds in award-winning computer graphics, optical illusions, and games that clarify memorable insights into self-similarity. 1992 edition. 448pp. 6 1/8 x 9 1/4. 0-486-47204-3

SET THEORY AND THE CONTINUUM PROBLEM, Raymond M. Smullyan and Melvin Fitting. A lucid, elegant, and complete survey of set theory, this three-part treatment explores axiomatic set theory, the consistency of the continuum hypothesis, and forcing and independence results. 1996 edition. 336pp. 6 x 9. 0-486-47484-4

DYNAMICAL SYSTEMS, Shlomo Sternberg. A pioneer in the field of dynamical systems discusses one-dimensional dynamics, differential equations, random walks, iterated function systems, symbolic dynamics, and Markov chains. Supplementary materials include PowerPoint slides and MATLAB exercises. 2010 edition. 272pp. 6 1/8 x 9 1/4. 0-486-47705-3

ORDINARY DIFFERENTIAL EQUATIONS, Morris Tenenbaum and Harry Pollard. Skillfully organized introductory text examines origin of differential equations, then defines basic terms and outlines general solution of a differential equation. Explores integrating factors; dilution and accretion problems; Laplace Transforms; Newton's Interpolation Formulas, more. 818pp. 5 3/8 x 8 1/2. 0-486-64940-7

MATROID THEORY, D. J. A. Welsh. Text by a noted expert describes standard examples and investigation results, using elementary proofs to develop basic matroid properties before advancing to a more sophisticated treatment. Includes numerous exercises. 1976 edition. 448pp. 5 3/8 x 8 1/2. 0-486-47439-9

THE CONCEPT OF A RIEMANN SURFACE, Hermann Weyl. This classic on the general history of functions combines function theory and geometry, forming the basis of the modern approach to analysis, geometry, and topology. 1955 edition. 208pp. 5 3/8 x 8 1/2. 0-486-47004-0

THE LAPLACE TRANSFORM, David Vernon Widder. This volume focuses on the Laplace and Stieltjes transforms, offering a highly theoretical treatment. Topics include fundamental formulas, the moment problem, monotonic functions, and Tauberian theorems. 1941 edition. 416pp. 5 3/8 x 8 1/2. 0-486-47755-X

Browse over 9,000 books at www.doverpublications.com

CATALOG OF DOVER BOOKS

Mathematics-History

THE WORKS OF ARCHIMEDES, Archimedes. Translated by Sir Thomas Heath. Complete works of ancient geometer feature such topics as the famous problems of the ratio of the areas of a cylinder and an inscribed sphere; the properties of conoids, spheroids, and spirals; more. 326pp. 5 3/8 x 8 1/2. 0-486-42084-1

THE HISTORICAL ROOTS OF ELEMENTARY MATHEMATICS, Lucas N. H. Bunt, Phillip S. Jones, and Jack D. Bedient. Exciting, hands-on approach to understanding fundamental underpinnings of modern arithmetic, algebra, geometry and number systems examines their origins in early Egyptian, Babylonian, and Greek sources. 336pp. 5 3/8 x 8 1/2. 0-486-25563-8

THE THIRTEEN BOOKS OF EUCLID'S ELEMENTS, Euclid. Contains complete English text of all 13 books of the Elements plus critical apparatus analyzing each definition, postulate, and proposition in great detail. Covers textual and linguistic matters; mathematical analyses of Euclid's ideas; classical, medieval, Renaissance and modern commentators; refutations, supports, extrapolations, reinterpretations and historical notes. 995 figures. Total of 1,425pp. All books 5 3/8 x 8 1/2.
Vol. I: 443pp. 0-486-60088-2
Vol. II: 464pp. 0-486-60089-0
Vol. III: 546pp. 0-486-60090-4

A HISTORY OF GREEK MATHEMATICS, Sir Thomas Heath. This authoritative two-volume set that covers the essentials of mathematics and features every landmark innovation and every important figure, including Euclid, Apollonius, and others. 5 3/8 x 8 1/2.
Vol. I: 461pp. 0-486-24073-8
Vol. II: 597pp. 0-486-24074-6

A MANUAL OF GREEK MATHEMATICS, Sir Thomas L. Heath. This concise but thorough history encompasses the enduring contributions of the ancient Greek mathematicians whose works form the basis of most modern mathematics. Discusses Pythagorean arithmetic, Plato, Euclid, more. 1931 edition. 576pp. 5 3/8 x 8 1/2.
0-486-43231-9

CHINESE MATHEMATICS IN THE THIRTEENTH CENTURY, Ulrich Libbrecht. An exploration of the 13th-century mathematician Ch'in, this fascinating book combines what is known of the mathematician's life with a history of his only extant work, the Shu-shu chiu-chang. 1973 edition. 592pp. 5 3/8 x 8 1/2.
0-486-44619-0

PHILOSOPHY OF MATHEMATICS AND DEDUCTIVE STRUCTURE IN EUCLID'S ELEMENTS, Ian Mueller. This text provides an understanding of the classical Greek conception of mathematics as expressed in Euclid's Elements. It focuses on philosophical, foundational, and logical questions and features helpful appendixes. 400pp. 6 1/2 x 9 1/4. 0-486-45300-6

BEYOND GEOMETRY: Classic Papers from Riemann to Einstein, Edited with an Introduction and Notes by Peter Pesic. This is the only English-language collection of these 8 accessible essays. They trace seminal ideas about the foundations of geometry that led to Einstein's general theory of relativity. 224pp. 6 1/8 x 9 1/4. 0-486-45350-2

HISTORY OF MATHEMATICS, David E. Smith. Two-volume history – from Egyptian papyri and medieval maps to modern graphs and diagrams. Non-technical chronological survey with thousands of biographical notes, critical evaluations, and contemporary opinions on over 1,100 mathematicians. 5 3/8 x 8 1/2.
Vol. I: 618pp. 0-486-20429-4
Vol. II: 736pp. 0-486-20430-8

Browse over 9,000 books at www.doverpublications.com

CATALOG OF DOVER BOOKS

Mathematics-Logic and Problem Solving

PERPLEXING PUZZLES AND TANTALIZING TEASERS, Martin Gardner. Ninety-three riddles, mazes, illusions, tricky questions, word and picture puzzles, and other challenges offer hours of entertainment for youngsters. Filled with rib-tickling drawings. Solutions. 224pp. 5 3/8 x 8 1/2. 0-486-25637-5

MY BEST MATHEMATICAL AND LOGIC PUZZLES, Martin Gardner. The noted expert selects 70 of his favorite "short" puzzles. Includes The Returning Explorer, The Mutilated Chessboard, Scrambled Box Tops, and dozens more. Complete solutions included. 96pp. 5 3/8 x 8 1/2. 0-486-28152-3

THE LADY OR THE TIGER?: and Other Logic Puzzles, Raymond M. Smullyan. Created by a renowned puzzle master, these whimsically themed challenges involve paradoxes about probability, time, and change; metapuzzles; and self-referentiality. Nineteen chapters advance in difficulty from relatively simple to highly complex. 1982 edition. 240pp. 5 3/8 x 8 1/2. 0-486-47027-X

SATAN, CANTOR AND INFINITY: Mind-Boggling Puzzles, Raymond M. Smullyan. A renowned mathematician tells stories of knights and knaves in an entertaining look at the logical precepts behind infinity, probability, time, and change. Requires a strong background in mathematics. Complete solutions. 288pp. 5 3/8 x 8 1/2.
0-486-47036-9

THE RED BOOK OF MATHEMATICAL PROBLEMS, Kenneth S. Williams and Kenneth Hardy. Handy compilation of 100 practice problems, hints and solutions indispensable for students preparing for the William Lowell Putnam and other mathematical competitions. Preface to the First Edition. Sources. 1988 edition. 192pp. 5 3/8 x 8 1/2. 0-486-69415-1

KING ARTHUR IN SEARCH OF HIS DOG AND OTHER CURIOUS PUZZLES, Raymond M. Smullyan. This fanciful, original collection for readers of all ages features arithmetic puzzles, logic problems related to crime detection, and logic and arithmetic puzzles involving King Arthur and his Dogs of the Round Table. 160pp. 5 3/8 x 8 1/2.
0-486-47435-6

UNDECIDABLE THEORIES: Studies in Logic and the Foundation of Mathematics, Alfred Tarski in collaboration with Andrzej Mostowski and Raphael M. Robinson. This well-known book by the famed logician consists of three treatises: "A General Method in Proofs of Undecidability," "Undecidability and Essential Undecidability in Mathematics," and "Undecidability of the Elementary Theory of Groups." 1953 edition. 112pp. 5 3/8 x 8 1/2. 0-486-47703-7

LOGIC FOR MATHEMATICIANS, J. Barkley Rosser. Examination of essential topics and theorems assumes no background in logic. "Undoubtedly a major addition to the literature of mathematical logic." – *Bulletin of the American Mathematical Society*. 1978 edition. 592pp. 6 1/8 x 9 1/4. 0-486-46898-4

INTRODUCTION TO PROOF IN ABSTRACT MATHEMATICS, Andrew Wohlgemuth. This undergraduate text teaches students what constitutes an acceptable proof, and it develops their ability to do proofs of routine problems as well as those requiring creative insights. 1990 edition. 384pp. 6 1/2 x 9 1/4. 0-486-47854-8

FIRST COURSE IN MATHEMATICAL LOGIC, Patrick Suppes and Shirley Hill. Rigorous introduction is simple enough in presentation and context for wide range of students. Symbolizing sentences; logical inference; truth and validity; truth tables; terms, predicates, universal quantifiers; universal specification and laws of identity; more. 288pp. 5 3/8 x 8 1/2. 0-486-42259-3

Browse over 9,000 books at www.doverpublications.com

CATALOG OF DOVER BOOKS

Physics

THEORETICAL NUCLEAR PHYSICS, John M. Blatt and Victor F. Weisskopf. An uncommonly clear and cogent investigation and correlation of key aspects of theoretical nuclear physics by leading experts: the nucleus, nuclear forces, nuclear spectroscopy, two-, three- and four-body problems, nuclear reactions, beta-decay and nuclear shell structure. 896pp. 5 3/8 x 8 1/2. 0-486-66827-4

QUANTUM THEORY, David Bohm. This advanced undergraduate-level text presents the quantum theory in terms of qualitative and imaginative concepts, followed by specific applications worked out in mathematical detail. 655pp. 5 3/8 x 8 1/2. 0-486-65969-0

ATOMIC PHYSICS AND HUMAN KNOWLEDGE, Niels Bohr. Articles and speeches by the Nobel Prize–winning physicist, dating from 1934 to 1958, offer philosophical explorations of the relevance of atomic physics to many areas of human endeavor. 1961 edition. 112pp. 5 3/8 x 8 1/2. 0-486-47928-5

COSMOLOGY, Hermann Bondi. A co-developer of the steady-state theory explores his conception of the expanding universe. This historic book was among the first to present cosmology as a separate branch of physics. 1961 edition. 192pp. 5 3/8 x 8 1/2. 0-486-47483-6

LECTURES ON QUANTUM MECHANICS, Paul A. M. Dirac. Four concise, brilliant lectures on mathematical methods in quantum mechanics from Nobel Prize-winning quantum pioneer build on idea of visualizing quantum theory through the use of classical mechanics. 96pp. 5 3/8 x 8 1/2. 0-486-41713-1

THE PRINCIPLE OF RELATIVITY, Albert Einstein and Frances A. Davis. Eleven papers that forged the general and special theories of relativity include seven papers by Einstein, two by Lorentz, and one each by Minkowski and Weyl. 1923 edition. 240pp. 5 3/8 x 8 1/2. 0-486-60081-5

PHYSICS OF WAVES, William C. Elmore and Mark A. Heald. Ideal as a classroom text or for individual study, this unique one-volume overview of classical wave theory covers wave phenomena of acoustics, optics, electromagnetic radiations, and more. 477pp. 5 3/8 x 8 1/2. 0-486-64926-1

THERMODYNAMICS, Enrico Fermi. In this classic of modern science, the Nobel Laureate presents a clear treatment of systems, the First and Second Laws of Thermodynamics, entropy, thermodynamic potentials, and much more. Calculus required. 160pp. 5 3/8 x 8 1/2. 0-486-60361-X

QUANTUM THEORY OF MANY-PARTICLE SYSTEMS, Alexander L. Fetter and John Dirk Walecka. Self-contained treatment of nonrelativistic many-particle systems discusses both formalism and applications in terms of ground-state (zero-temperature) formalism, finite-temperature formalism, canonical transformations, and applications to physical systems. 1971 edition. 640pp. 5 3/8 x 8 1/2. 0-486-42827-3

QUANTUM MECHANICS AND PATH INTEGRALS: Emended Edition, Richard P. Feynman and Albert R. Hibbs. Emended by Daniel F. Styer. The Nobel Prize–winning physicist presents unique insights into his theory and its applications. Feynman starts with fundamentals and advances to the perturbation method, quantum electrodynamics, and statistical mechanics. 1965 edition, emended in 2005. 384pp. 6 1/8 x 9 1/4. 0-486-47722-3

Browse over 9,000 books at www.doverpublications.com

CATALOG OF DOVER BOOKS

Physics

INTRODUCTION TO MODERN OPTICS, Grant R. Fowles. A complete basic undergraduate course in modern optics for students in physics, technology, and engineering. The first half deals with classical physical optics; the second, quantum nature of light. Solutions. 336pp. 5 3/8 x 8 1/2. 0-486-65957-7

THE QUANTUM THEORY OF RADIATION: Third Edition, W. Heitler. The first comprehensive treatment of quantum physics in any language, this classic introduction to basic theory remains highly recommended and widely used, both as a text and as a reference. 1954 edition. 464pp. 5 3/8 x 8 1/2. 0-486-64558-4

QUANTUM FIELD THEORY, Claude Itzykson and Jean-Bernard Zuber. This comprehensive text begins with the standard quantization of electrodynamics and perturbative renormalization, advancing to functional methods, relativistic bound states, broken symmetries, nonabelian gauge fields, and asymptotic behavior. 1980 edition. 752pp. 6 1/2 x 9 1/4. 0-486-44568-2

FOUNDATIONS OF POTENTIAL THERY, Oliver D. Kellogg. Introduction to fundamentals of potential functions covers the force of gravity, fields of force, potentials, harmonic functions, electric images and Green's function, sequences of harmonic functions, fundamental existence theorems, and much more. 400pp. 5 3/8 x 8 1/2.
0-486-60144-7

FUNDAMENTALS OF MATHEMATICAL PHYSICS, Edgar A. Kraut. Indispensable for students of modern physics, this text provides the necessary background in mathematics to study the concepts of electromagnetic theory and quantum mechanics. 1967 edition. 480pp. 6 1/2 x 9 1/4. 0-486-45809-1

GEOMETRY AND LIGHT: The Science of Invisibility, Ulf Leonhardt and Thomas Philbin. Suitable for advanced undergraduate and graduate students of engineering, physics, and mathematics and scientific researchers of all types, this is the first authoritative text on invisibility and the science behind it. More than 100 full-color illustrations, plus exercises with solutions. 2010 edition. 288pp. 7 x 9 1/4. 0-486-47693-6

QUANTUM MECHANICS: New Approaches to Selected Topics, Harry J. Lipkin. Acclaimed as "excellent" (*Nature*) and "very original and refreshing" (*Physics Today*), these studies examine the Mössbauer effect, many-body quantum mechanics, scattering theory, Feynman diagrams, and relativistic quantum mechanics. 1973 edition. 480pp. 5 3/8 x 8 1/2. 0-486-45893-8

THEORY OF HEAT, James Clerk Maxwell. This classic sets forth the fundamentals of thermodynamics and kinetic theory simply enough to be understood by beginners, yet with enough subtlety to appeal to more advanced readers, too. 352pp. 5 3/8 x 8 1/2. 0-486-41735-2

QUANTUM MECHANICS, Albert Messiah. Subjects include formalism and its interpretation, analysis of simple systems, symmetries and invariance, methods of approximation, elements of relativistic quantum mechanics, much more. "Strongly recommended." – *American Journal of Physics*. 1152pp. 5 3/8 x 8 1/2. 0-486-40924-4

RELATIVISTIC QUANTUM FIELDS, Charles Nash. This graduate-level text contains techniques for performing calculations in quantum field theory. It focuses chiefly on the dimensional method and the renormalization group methods. Additional topics include functional integration and differentiation. 1978 edition. 240pp. 5 3/8 x 8 1/2.
0-486-47752-5

Browse over 9,000 books at www.doverpublications.com

Physics

MATHEMATICAL TOOLS FOR PHYSICS, James Nearing. Encouraging students' development of intuition, this original work begins with a review of basic mathematics and advances to infinite series, complex algebra, differential equations, Fourier series, and more. 2010 edition. 496pp. 6 1/8 x 9 1/4. 0-486-48212-X

TREATISE ON THERMODYNAMICS, Max Planck. Great classic, still one of the best introductions to thermodynamics. Fundamentals, first and second principles of thermodynamics, applications to special states of equilibrium, more. Numerous worked examples. 1917 edition. 297pp. 5 3/8 x 8. 0-486-66371-X

AN INTRODUCTION TO RELATIVISTIC QUANTUM FIELD THEORY, Silvan S. Schweber. Complete, systematic, and self-contained, this text introduces modern quantum field theory. "Combines thorough knowledge with a high degree of didactic ability and a delightful style." – *Mathematical Reviews*. 1961 edition. 928pp. 5 3/8 x 8 1/2. 0-486-44228-4

THE ELECTROMAGNETIC FIELD, Albert Shadowitz. Comprehensive undergraduate text covers basics of electric and magnetic fields, building up to electromagnetic theory. Related topics include relativity theory. Over 900 problems, some with solutions. 1975 edition. 768pp. 5 5/8 x 8 1/4. 0-486-65660-8

THE PRINCIPLES OF STATISTICAL MECHANICS, Richard C. Tolman. Definitive treatise offers a concise exposition of classical statistical mechanics and a thorough elucidation of quantum statistical mechanics, plus applications of statistical mechanics to thermodynamic behavior. 1930 edition. 704pp. 5 5/8 x 8 1/4.
0-486-63896-0

INTRODUCTION TO THE PHYSICS OF FLUIDS AND SOLIDS, James S. Trefil. This interesting, informative survey by a well-known science author ranges from classical physics and geophysical topics, from the rings of Saturn and the rotation of the galaxy to underground nuclear tests. 1975 edition. 320pp. 5 3/8 x 8 1/2. 0-486-47437-2

STATISTICAL PHYSICS, Gregory H. Wannier. Classic text combines thermodynamics, statistical mechanics, and kinetic theory in one unified presentation. Topics include equilibrium statistics of special systems, kinetic theory, transport coefficients, and fluctuations. Problems with solutions. 1966 edition. 532pp. 5 3/8 x 8 1/2.
0-486-65401-X

SPACE, TIME, MATTER, Hermann Weyl. Excellent introduction probes deeply into Euclidean space, Riemann's space, Einstein's general relativity, gravitational waves and energy, and laws of conservation. "A classic of physics." – *British Journal for Philosophy and Science*. 330pp. 5 3/8 x 8 1/2. 0-486-60267-2

RANDOM VIBRATIONS: Theory and Practice, Paul H. Wirsching, Thomas L. Paez and Keith Ortiz. Comprehensive text and reference covers topics in probability, statistics, and random processes, plus methods for analyzing and controlling random vibrations. Suitable for graduate students and mechanical, structural, and aerospace engineers. 1995 edition. 464pp. 5 3/8 x 8 1/2. 0-486-45015-5

PHYSICS OF SHOCK WAVES AND HIGH-TEMPERATURE HYDRO DYNAMIC PHENOMENA, Ya B. Zel'dovich and Yu P. Raizer. Physical, chemical processes in gases at high temperatures are focus of outstanding text, which combines material from gas dynamics, shock-wave theory, thermodynamics and statistical physics, other fields. 284 illustrations. 1966–1967 edition. 944pp. 6 1/8 x 9 1/4.
0-486-42002-7

Browse over 9,000 books at www.doverpublications.com

CATALOG OF DOVER BOOKS

Engineering

FUNDAMENTALS OF ASTRODYNAMICS, Roger R. Bate, Donald D. Mueller, and Jerry E. White. Teaching text developed by U.S. Air Force Academy develops the basic two-body and n-body equations of motion; orbit determination; classical orbital elements, coordinate transformations; differential correction; more. 1971 edition. 455pp. 5 3/8 x 8 1/2. 0-486-60061-0

INTRODUCTION TO CONTINUUM MECHANICS FOR ENGINEERS: Revised Edition, Ray M. Bowen. This self-contained text introduces classical continuum models within a modern framework. Its numerous exercises illustrate the governing principles, linearizations, and other approximations that constitute classical continuum models. 2007 edition. 320pp. 6 1/8 x 9 1/4. 0-486-47460-7

ENGINEERING MECHANICS FOR STRUCTURES, Louis L. Bucciarelli. This text explores the mechanics of solids and statics as well as the strength of materials and elasticity theory. Its many design exercises encourage creative initiative and systems thinking. 2009 edition. 320pp. 6 1/8 x 9 1/4. 0-486-46855-0

FEEDBACK CONTROL THEORY, John C. Doyle, Bruce A. Francis and Allen R. Tannenbaum. This excellent introduction to feedback control system design offers a theoretical approach that captures the essential issues and can be applied to a wide range of practical problems. 1992 edition. 224pp. 6 1/2 x 9 1/4. 0-486-46933-6

THE FORCES OF MATTER, Michael Faraday. These lectures by a famous inventor offer an easy-to-understand introduction to the interactions of the universe's physical forces. Six essays explore gravitation, cohesion, chemical affinity, heat, magnetism, and electricity. 1993 edition. 96pp. 5 3/8 x 8 1/2. 0-486-47482-8

DYNAMICS, Lawrence E. Goodman and William H. Warner. Beginning engineering text introduces calculus of vectors, particle motion, dynamics of particle systems and plane rigid bodies, technical applications in plane motions, and more. Exercises and answers in every chapter. 619pp. 5 3/8 x 8 1/2. 0-486-42006-X

ADAPTIVE FILTERING PREDICTION AND CONTROL, Graham C. Goodwin and Kwai Sang Sin. This unified survey focuses on linear discrete-time systems and explores natural extensions to nonlinear systems. It emphasizes discrete-time systems, summarizing theoretical and practical aspects of a large class of adaptive algorithms. 1984 edition. 560pp. 6 1/2 x 9 1/4. 0-486-46932-8

INDUCTANCE CALCULATIONS, Frederick W. Grover. This authoritative reference enables the design of virtually every type of inductor. It features a single simple formula for each type of inductor, together with tables containing essential numerical factors. 1946 edition. 304pp. 5 3/8 x 8 1/2. 0-486-47440-2

THERMODYNAMICS: Foundations and Applications, Elias P. Gyftopoulos and Gian Paolo Beretta. Designed by two MIT professors, this authoritative text discusses basic concepts and applications in detail, emphasizing generality, definitions, and logical consistency. More than 300 solved problems cover realistic energy systems and processes. 800pp. 6 1/8 x 9 1/4. 0-486-43932-1

THE FINITE ELEMENT METHOD: Linear Static and Dynamic Finite Element Analysis, Thomas J. R. Hughes. Text for students without in-depth mathematical training, this text includes a comprehensive presentation and analysis of algorithms of time-dependent phenomena plus beam, plate, and shell theories. Solution guide available upon request. 672pp. 6 1/2 x 9 1/4. 0-486-41181-8

Browse over 9,000 books at www.doverpublications.com